Medical-Grade S Development

This book is a practical guide to meeting IEC 62304 software-development requirements within the context of an ISO 13485 quality management system (QMS). The book proves this can be done with a minimum amount of friction, overlap, and back-and-forth between development stages. It essentially shows you how you should shape your medical-software development processes to fit in with the QMS processes in the smartest and leanest way possible.

By following the advice in this book, you can reuse processes from your QMS, ensure your product-realization processes meet the requirements for medical-software development, and marry all the requirements together using tried and tested solutions into one efficient system. The expertise of the authors here goes beyond just the experiences of one real-world project as they tap into over 30 years of experience and countless software and software-assessment projects to distill their advice.

The book takes a hands-on approach by first teaching you the top 25 lessons to know before starting to develop a process for medical-software development. It then walks you through the expectations placed on the key aspects of such a process by the key standards. The book progresses from an overview of both standards and the general requirements involved to a detailed discussion of the expected stages from software development and maintenance to risk management, configuration management, and problem resolution. The book provides insightful advice on how the requirements of the IEC 62304 software-development life cycle can be married with an ISO 13485 QMS, how the development of the technical file should be organized, and how to address conformity assessment, the daily after-approval, and the recent trends that will affect the industry in the coming years.

The book is modeled after the IEC 62304 standard and adopts its clause structure in the numbering of sections for easy reference. The book does not attempt to replicate either standard. For the ISO 13485 standard, it recites the necessary requirements succinctly. For IEC 62304, the discussion is in-depth and also addresses the impact of ISO 13485 on the requirements discussed. In this way, the book drills into both standards to expose the core of each requirement and shape these into a practical, cohesive workflow for developing, maintaining, and improving a Lean software development pipeline.

Medical-Grade Software Development

How to Build Medical-Device Products That Meet the Requirements of IEC 62304 and ISO 13485

Ilkka Juuso
Ilpo Pöyhönen

A PRODUCTIVITY PRESS BOOK

First published 2024
by Routledge
605 Third Avenue, New York, NY 10158

and by Routledge
4 Park Square, Milton Park, Abingdon, Oxon, OX14 4RN

Routledge is an imprint of the Taylor & Francis Group, an informa business

© 2024 Ilkka Juuso & Ilpo Pöyhönen

The right of Ilkka Juuso & Ilpo Pöyhönen to be identified as author of this work has been asserted by them in accordance with sections 77 and 78 of the Copyright, Designs and Patents Act 1988.

All rights reserved. No part of this book may be reprinted or reproduced or utilised in any form or by any electronic, mechanical, or other means, now known or hereafter invented, including photocopying and recording, or in any information storage or retrieval system, without permission in writing from the publishers.

Trademark notice: Product or corporate names may be trademarks or registered trademarks, and are used only for identification and explanation without intent to infringe.

ISBN: 9781032593180 (hbk)
ISBN: 9781032594729 (pbk)
ISBN: 9781003454830 (ebk)

DOI: 10.4324/9781003454830

Typeset in Garamond
by Deanta Global Publishing Services, Chennai, India

Dedicated to encounters of the best kind
– and learning from it all.

Contents

List of Figures and Tables xiv
Preface xix
Acknowledgments xxiv
About the Authors xxvi

1 **What to Know before Getting Started** 1
   1.1 There Is No One Correct Answer to Anything 2
   1.2 Why Do We Need a Standard at All? 5
   1.3 What Is Medical-Grade Software? 8
   1.4 Software Is Classified by the Safety Concerns Associated with It .. 10
   1.5 Software Is Made of Units, Items, and Systems 12
   1.6 Software Development Has a Life Cycle 14
   1.7 Everything Is a Process, and Those Consist of Activities and Tasks 16
   1.8 IEC 62304 Does Not Insist on a Specific Process Structure 18
   1.9 IEC 62304 Does Not Impose a Documentation Set 19
   1.10 Applicable Regulations as Part of the Requirements 20
   1.11 Building on the Quagmire That Surrounds IEC 62304 22
   1.12 Retaining Trust at Each Step of the Way Is Perhaps the Most Critical of Tasks 24
   1.13 The Relationship between Code and Documentation is at the Core of the Marriage, Even after a Shotgun-Wedding 26
   1.14 Evaluation, Review, Verification, and Validation 27
   1.15 How Many Reviews Is Enough? 31
   1.16 How Much Testing Is Enough? 34
   1.17 Fail and Pass 38
   1.18 One Step Forward, Two Steps Back 41
   1.19 Continuous Improvement 43

1.20 What Is SOUP? 45
1.21 Configuration Items and Controlled Changes 46
1.22 Legacy Software Has a Dedicated Pathway Through IEC 62304...48
1.23 Software Always Fails, and Other Software Risk Beliefs................49
1.24 Certification Bodies and the Notified Body 51
1.25 Safety Is Relative 53

2 ISO 13485 as the Backbone of It All 55
2.1 The Whole of the Quality Management System 56
2.2 The Process for Product Realization...57

3 IEC 62304 as The Flesh around The Bones 61
3.1 The Past, Present, and Future...62
    3.1.1 Times of the Primordial Ooze...62
    3.1.2 Modern Times 65
    3.1.3 The Future 67
3.2 Contents...69
    3.2.1 Foreword...69
    3.2.2 Introduction..70
    3.2.3 Clause 1: Scope..73
    3.2.4 Clause 2: Normative References 76
    3.2.5 Clause 3: Terms and Definitions 77
    3.2.6 Clause 4: General Requirements 79
    3.2.7 Clause 5: Software Development Process.........................80
    3.2.8 Clause 6: Software Maintenance Process..........................80
    3.2.9 Clause 7: Software Risk-Management Process 82
    3.2.10 Clause 8: Software Configuration Management Process......82
    3.2.11 Clause 9: Software Problem Resolution Process..............83
    3.2.12 Annex A – Rationale for the Requirements of This Standard 84
    3.2.13 Annex B – Guidance on the Provisions of This Standard ...85
    3.2.14 Annex C – Relationship to Other Standards.......................85
    3.2.15 Annex D – Implementation...88

4 General Requirements 89
4.1 Quality Management Systems...90
4.2 Risk Management 91
4.3 Software Safety Classification 92
    4.3.1 Classification Formula..93
    4.3.2 What Is Harm? 94
    4.3.3 Accept That All Software Will Fail 95

		4.3.4	Use of External Risk Controls 96
		4.3.5	Reducing Both Probability and Severity 97
		4.3.6	Recording and Inheriting the Classification 98
		4.3.7	Final Remarks 99
	4.4	Legacy Software 100
		4.4.1	Risk-Management Activities 100
		4.4.2	Gap Analysis 101
		4.4.3	Gap Closure Activities 103
		4.4.4	Rationale for Use of Legacy Software 104

5 Software Development 105
	5.1	Development Planning 108
		5.1.1	Expectations from ISO 13485 109
		5.1.2	Expectations from IEC 62304 111
		5.1.3	Suggested Synthesis 116
	5.2	Requirements Analysis 124
		5.2.1	Expectations from ISO 13485 126
		5.2.2	Expectations from IEC 62304 127
		5.2.3	Suggested Synthesis 130
	5.3	Architectural Design 132
		5.3.1	Expectations from ISO 13485 133
		5.3.2	Expectations from IEC 62304 134
		5.3.3	Suggested Synthesis 137
	5.4	Detailed Design 138
		5.4.1	Expectations from ISO 13485 139
		5.4.2	Expectations from IEC 62304 140
		5.4.3	Suggested Synthesis 141
	5.5	Unit Implementation and Verification 141
		5.5.1	Expectations from ISO 13485 142
		5.5.2	Expectations from IEC 62304 143
		5.5.3	Suggested Synthesis 145
	5.6	Integration and Integration Testing 146
		5.6.1	Expectations from ISO 13485 147
		5.6.2	Expectations from IEC 62304 148
		5.6.3	Suggested Synthesis 151
	5.7	Software-System Testing 152
		5.7.1	Expectations from ISO 13485 153
		5.7.2	Expectations from IEC 62304 154
		5.7.3	Suggested Synthesis 156

5.8　Release 156
　　　　5.8.1　Expectations from ISO 13485 158
　　　　5.8.2　Expectations from IEC 62304 162
　　　　5.8.3　Suggested Synthesis 165
　　5.9　The Parts Left Out by IEC 62304 170
　　　　5.9.1　Conducting Reviews in D&D Stages 170
　　　　5.9.2　D&D Verification 170
　　　　5.9.3　D&D Validation 172

6　Software Maintenance 175
　　6.1　Software Maintenance Plan 176
　　　　6.1.1　Expectations from ISO 13485 177
　　　　6.1.2　Expectations from IEC 62304 179
　　　　6.1.3　Suggested Synthesis 179
　　6.2　Problem and Modification Analysis 180
　　　　6.2.1　Expectations from ISO 13485 182
　　　　6.2.2　Expectations from IEC 62304 182
　　　　6.2.3　Suggested Synthesis 184
　　6.3　Modification Implementation 185
　　　　6.3.1　Expectations from ISO 13485 186
　　　　6.3.2　Expectations from IEC 62304 186
　　　　6.3.3　Suggested Synthesis 187
　　6.4　The Parts Left Out By IEC 62304 188

7　Risk Management 189
　　7.1　Analysis of Software Contributing to Hazardous Situations 194
　　　　7.1.1　Expectations from ISO 13485 197
　　　　7.1.2　Expectations from IEC 62304 197
　　　　7.1.3　Suggested Synthesis 199
　　7.2　Risk-Control Measures 199
　　　　7.2.1　Expectations from ISO 13485 199
　　　　7.2.2　Expectations from IEC 62304 199
　　　　7.2.3　Suggested Synthesis 200
　　7.3　Verification of Risk-Control Measures 200
　　　　7.3.1　Expectations from ISO 13485 200
　　　　7.3.2　Expectations from IEC 62304 200
　　　　7.3.3　Suggested Synthesis 201
　　7.4　Risk Management of Software Changes 201
　　　　7.4.1　Expectations from ISO 13485 201

		7.4.2	Expectations from IEC 62304	202

  7.4.2 Expectations from IEC 62304 202
  7.4.3 Suggested Synthesis 202
 7.5 The Parts Left Out by IEC 62304 202

8 Configuration Management 205
 8.1 Identification of Configuration Items...207
  8.1.1 Expectations from ISO 13485 208
  8.1.2 Expectations from IEC 62304 208
  8.1.3 Suggested Synthesis 209
 8.2 Change Control (Incl. Verification) 215
  8.2.1 Expectations from ISO 13485 215
  8.2.2 Expectations from IEC 62304 216
  8.2.3 Suggested Synthesis 217
 8.3 History of Controlled Items 218
  8.3.1 Expectations from ISO 13485 219
  8.3.2 Expectations from IEC 62304 219
  8.3.3 Suggested Synthesis 219
 8.4 The Parts Left Out by IEC 62304 220

9 Problem Resolution 221
 9.1 Prepare Problem Reports 222
  9.1.1 Expectations from ISO 13485 224
  9.1.2 Expectations from IEC 62304 225
  9.1.3 Suggested Synthesis 225
 9.2 Investigate the Problem 226
  9.2.1 Expectations from ISO 13485 227
  9.2.2 Expectations from IEC 62304 227
  9.2.3 Suggested Synthesis 228
 9.3 Advise Relevant Parties 229
  9.3.1 Expectations from ISO 13485 230
  9.3.2 Expectations from IEC 62304 230
  9.3.3 Suggested Synthesis 230
 9.4 Use Change-Control Processes 231
  9.4.1 Expectations from ISO 13485 232
  9.4.2 Expectations from IEC 62304 232
  9.4.3 Suggested Synthesis 232
 9.5 Maintain Records 233
  9.5.1 Expectations from ISO 13485 234
  9.5.2 Expectations from IEC 62304 234
  9.5.3 Suggested Synthesis 234

9.6 Analyze Problems for Trends 235
    9.6.1 Expectations from ISO 13485 235
    9.6.2 Expectations from IEC 62304 236
    9.6.3 Suggested Synthesis 236
9.7 Verify Software Problem Resolution 237
    9.7.1 Expectations from ISO 13485 238
    9.7.2 Expectations from IEC 62304 239
    9.7.3 Suggested Synthesis 240
9.8 Test Documentation Contents 241
    9.8.1 Expectations from ISO 13485 241
    9.8.2 Expectations from IEC 62304 242
    9.8.3 Suggested Synthesis 242
9.9 The Parts Left Out by IEC 62304 243

10 Integration with Your QMS 245
10.1 Write Your Product Realization Processes in the QMS 247
10.2 Develop Your Planning Documentation According to Your QMS 250
10.3 Execute on Your Plans as Expected by Your Arrangements 250
10.4 Execute on Your Post Release Activities 251
10.5 Keep Your QMS in Good Shape 251

11 Technical Documentation 252
11.1 What Is Technical Documentation Anyway? 253
11.2 Process Documents 256
11.3 Development Records 256
11.4 The Audit Package 259

12 Seeing into the Future 260
12.1 Agile Software Development 261
    12.1.1 What Is Agile Anyway? 262
    12.1.2 Development Life Cycle in a Regulated Environment 265
12.2 IEC 82304-1 on Health Software 267
    12.2.1 Product Requirements 269
    12.2.2 Software Development Process 271
    12.2.3 Validation 271
    12.2.4 Product Documentation 272
    12.2.5 Post-Market Activities 275
12.3 Segregation in Software Architecture 276
12.4 Risk Management Influenced by FMEA 280

	12.5 Usability Engineering	284
	12.6 Cybersecurity	290
	12.6.1 Cybersecurity Over Product Life Cycle	291
	12.6.2 Cybersecurity Across Jurisdictions	293
	12.6.3 IEC 81001-5-1	294
	12.6.4 ISO/IEC 27001	295
	12.7 Artificial Intelligence	296
	12.8 Cloud Computing	300
	12.9 Edge Intelligence	302
13	**Conformity Assessment**	**304**
	13.1 Requirements Placed on the Assessor	305
	13.2 The Assessment	308
	13.3 Typical Shortcomings in Assessments	310
	13.4 What Happens Afterward?	313
14	**Regulatory Approval**	**315**
	14.1 Benefits of the IEC 62304 Test Report	315
	14.2 ISO 13485 Certification	317
	14.3 Medical-Device Conformity Assessment	318
	14.4 Changes	319
15	**Business as Usual**	**321**
	15.1 The Everyday Processes to Now Run	322
	15.2 The Future of IEC 62304	324
	15.3 The Future of ISO 13485	325
	15.4 The Joy of Compliance	325
16	**Conclusions**	**327**
References		**331**
Index		**332**

List of Figures and Tables

Figures

Figure 1.1 The sweet spot is wider than just a single hole to hit. 3

Figure 1.2 Standards are like maps to the otherwise uncharted world. 6

Figure 1.3 The definition of "medical software" is a slippery concept to nail down. 8

Figure 1.4 It's not quite as easy as in that nursery rhyme, but knowing your ABCs is required. 11

Figure 1.5 System begets an item, an item begets a unit. 12

Figure 1.6 Not to put too fine a point on it all, but a life cycle is just that. 15

Figure 1.7 Everything is a process. 17

Figure 1.8 The pieces of the puzzle are up to you. 18

Figure 1.9 Is a document by any other name not still a document? 19

Figure 1.10 Applicable regulations are a foundational part of your requirements. 20

Figure 1.11 The IEC 62304 has reached a point of stable uncertainty. 23

Figure 1.12 Steps to trust may be different in height, but the action is always the same. 24

Figure 1.13 Code and documentation should have a symbiotic relationship. 26

Figure 1.14 Verification and validation have similar goals, but different points of view. 28

Figure 1.15 "Review vertigo" may lead to more fatigue than attention to detail. 32

Figure 1.16 Testing is required to trust that it all works. 35

Figure 1.17 To pass or to fail, that is the Shakespearean question in software. 39

Figure 1.18 Maintaining forward momentum may at times be difficult....42

Figure 1.19 Your path into QMS should have a rising crescendo. 44

Figure 1.20 SOUP or not, you'll need to know what's in it. 45

Figure 1.21 Configuration covers both versions of your product and what is in it. 46

Figure 1.22 The IEC 62304 has a charted path to conformance for old software. 48

Figure 1.23 You don't have to be Tom Cruise to find your "Porsche" in a creek if you ignore the signs. 50

Figure 1.24 Know that not all certificates carry equal value. 51

Figure 1.25 Marshmallows on cobra fangs might increase your sense of safety, but will it be safe enough? 53

Figure 3.1 The collateral and particular standards of IEC 60601-1. 63

Figure 3.2 Scope of IEC 60601-1-4 on software. 64

Figure 3.3 Demonstration of compliance with standards IEC 60601-1 Clause 14, IEC 62304, and IEC 82304-1. 66

Figure 3.4 Word cloud for the Introduction. 74

Figure 3.5 Word cloud for Clause 1. 76

Figure 3.6 Word cloud for Clause 2. 77

Figure 3.7 Word cloud for Clause 3. 79

Figure 3.8 Word cloud for Clause 4. 80

Figure 3.9 Word cloud for Clause 5. 81

Figure 3.10 Word cloud for Clause 6. 82

Figure 3.11 Word cloud for Clause 7. 83

Figure 3.12 Word cloud for Clause 8. 83

Figure 3.13 Word cloud for Clause 9. 84

Figure 3.14 Word cloud for Annex A. 85

Figure 3.15 Word cloud for Annex B. 86

Figure 3.16 Word cloud for Annex C. 87

Figure 3.17 Word cloud for Annex D. 88

Figure 4.1 Key to safety classification. 93

Figure 4.2 The use of P1 and P2 in a risk matrix. 97

Figure 5.1 Both standards look to provide bookends and bookmarks for your work. 108

Figure 5.2 During planning your horizons are at their widest. 109

Figure 5.3 Requirements may be a bit of an octopus sushi to wrangle, but they should never be treated as an alphabet soup. 125

Figure 5.4 Floppies were once the modern equivalent to paper scrolls, but now they too may be relegated to ancient ruins on our architectural map. 133

Figure 5.5 Detailed design has us zooming into the architectural map we previously created. 138

Figure 5.6 After all that designing, this is where we move to first doing and then checking the results. 142

Figure 5.7 This is that "E pluribus unum" moment: out of many becomes one. 147

Figure 5.8 Testing should lead to that "all systems are go" feeling. 153

Figure 5.9 When the software is ready for the limelight, it's time. 158

Figure 6.1 Software maintenance must be on your roadmap. 177

Figure 6.2 My dog ate my software. 181

Figure 6.3 Problem-resolution process may be akin to field surgery. 185

Figure 7.1 Standalone networks, private networks, and the cloud each come with different risks associated with them. 196

Figure 11.1 Example structure for technical documentation. 255

Figure 11.2 Technical documentation in DIR and DOR. 257

Figure 11.3 Technical documentation in DTR. 257

Figure 12.1 Segregation breakdown due to memory access. 278

Figure 12.2 Hazards as a gateway to risk management. 281

Figure 12.3 RPN risk scores scaled to emphasize the role of severity. 283

Figure 12.4 Deriving use specifications and task descriptions from the intended use. 287

Figure 12.5 The difference between black-box and white-box AI.......... 298

Tables

Table 2.1 A Rough Mapping between ISO 13485 Clause 7 and IEC 62304 59

Table 5.1 Topics of Software Development 106

Table 5.2 Alignment of ISO 13485 and IEC 62304 Clause 5 107

Table 5.3 Synthesis of Top-Level Requirements 117

Table 5.4 Synthesis of Top-Level Requirements 131

Table 5.5 Synthesis of Top-Level Requirements 137

Table 5.6 Synthesis of Top-Level Requirements 141

Table 5.7 Synthesis of Top-Level Requirements 146

Table 5.8 Synthesis of Top-Level Requirements 152

Table 5.9 Synthesis of Top-Level Requirements 157

Table 5.10 Synthesis of Top-Level Requirements 166

Table 6.1 Topics of Software Maintenance 176

Table 6.2 Synthesis of Top-Level Requirements 180

Table 6.3 Synthesis of Top-Level Requirements 184

Table 6.4 Synthesis of Top-Level Requirements 187

Table 7.1	Topics of Risk Management	192
Table 8.1	Topics of Configuration Management	207
Table 8.2	Synthesis of Top-Level Requirements	209
Table 8.3	Synthesis of Top-Level Requirements	218
Table 8.4	Synthesis of Top-Level Requirements	219
Table 9.1	Topics of Problem Resolution	223
Table 9.2	Synthesis of Top-Level Requirements	226
Table 9.3	Synthesis of Top-Level Requirements	228
Table 9.4	Synthesis of Top-Level Requirements	231
Table 9.5	Synthesis of Top-Level Requirements	232
Table 9.6	Synthesis of Top-Level Requirements	234
Table 9.7	Synthesis of Top-Level Requirements	236
Table 9.8	Synthesis of Top-Level Requirements	240
Table 9.9	Synthesis of Top-Level Requirements	243
Table 10.1	Basic Software Development Life Cycle	248
Table 13.1	Excerpt From IEC 60601-1 Clause 14.6 Test Report	309
Table 13.2	Excerpt From IEC 62304 Clause 5.4 Test Report	309

Preface

Medical software has a lot to answer for. It is an industry based on decades of experience built up from the development of software in industries ranging from aviation and military to mobile applications and the piece of embedded binary DNA running your toaster. Software is in all these cases the logic or even the intelligence animating the physical devices, giving them an artificial brain or a rudimentary nerve cord of sorts. The structure and rigor involved in the development of software on these disparate domains, however, varies greatly. After all, the different application areas for software have vastly different characteristics. A mid-air software update to a jetliner on a transatlantic route is unacceptable and could cause the plane to crash. A forced mid-surgery operating system update in the hospital operating room will cause headlines across the world, but it does happen. An update to a popular mobile application or game on the other hand is expected and is often a sign of a well-maintained and evolving app.

Medical software has evolved out of this diverse pedigree. Some medical software today is developed by individual hobbyists, others by established medical-device companies, and yet others by big companies moving into health and wellbeing from other domains of business. The division between what is considered a simple health application and what is hardcore medical software is at times a judgment call, and occasionally similar, almost identical, software products can sit on opposite sides of this fence. At the same time, it is universally acknowledged that most advances in medical devices, or most new advanced features built into those devices, will require the use of software. Artificial intelligence (AI) is the popular umbrella term often used when talking about the unknowable next generation of medical devices. Some of the features covered by this umbrella will sit firmly in the domain of artificial intelligence, others will be content with applying more traditional machine learning technologies (ML), and yet others just invoke these terms generously to claim a marketing stake in the field. AI/ML will certainly usher in new generations of medical devices that perform increasingly advanced tasks on behalf of their human masters, and – even if we are not quite there yet – many eyes and much hope is placed on autonomous medical software.

The development of medical software in businesses unfolds in different ways, following different development models from waterfall to agile, but all of these will have to be linked to international standards and regulatory requirements before the resulting software can be sold and used. The IEC 62304 standard on software life-cycle management and the ISO 13485 standard on quality management are two of the cornerstones of medical-software development today. Most medical-device companies must abide by both to meet the expectations placed on them by their customers and regulators. These are not the only two standards you will perhaps want to consider – ISO 14971 risk management, IEC 62366-1 usability, and ISO 14155 clinical investigation will also inform your software operations in all likelihood – but these two are the foundational standards that will inform much of the fabric of your business in medical software.

Combining the IEC 62304 and ISO 13485 standards in a practical and practically extensible manner will be your second question to answer, right after, "How do I implement the correct quality management system for my company"? Appropriately, this book builds on the earlier book *Developing an ISO 13485-Certified Quality Management System – An Implementation Guide for the Medical-Device Industry* written by Dr Juuso (Routledge 2022). That earlier book focused on setting up a lean and powerful QMS at a medical-device company. The focus here in this book is on how to marry quality management and software development together in a practical way for producing medical software – whether that is software for use inside medical devices, software as a medical device, or something in between. If you are building a company from scratch, we recommend you first read the above-mentioned QMS book and then continue to the software-specific discussion with this book. If your interest is primarily in software development, and you won't be responsible for the rest of your QMS, you can jump ahead to here – the QMS book offers good context for the work here, but it may not be a must-read for you here and now. We also expect this book to be a handy tool for software suppliers who don't develop medical-device software themselves, but who are asked to meet some subset of the requirements by their B2B customers who do.

The discussion here adopts the structure of ISO 13485 quality management as its backbone and then builds on this to construct a practical framework for medical-grade software development. ISO 13485 is the big, beautiful standard your medical-device operations will be measured against, and against which you will ultimately want to retain your status and any possible official certification status. IEC 62304, by comparison, is a vastly popular supporting standard in medical-software development and later maintenance. It is, however, plagued by a strained, drawn-out revision process and has had to contend with gathering storm clouds since 2006. IEC 62304 is now technically stable for the next few years, but its future is contentious and may see major revision taking place. For this reason, it makes all the sense in the world to build your software-development efforts based on the bedrock offered by ISO 13485 and then add to this as IEC 62304 suggests.

This book was borne out of a wish to use ISO 13485 quality management as a diving board into medical-grade software development and to find answers to the grave questions posed to IEC 62304 over the past decade and a half. Furthermore, we wanted to provide practical solutions to developing medical software and the associated documentation in a smart, lean, and organically developer-friendly way.

Our whole philosophy in writing this book is to build a sturdy, but lightweight process for medical-software development. A process that ensures you meet your requirements in a lean, timely, straightforward, and safe manner. The process is based on our own experiences in designing, analyzing, and assessing medical-software development processes for medical-device companies. This work has been informed by the standard, the decades-long revision attempts around it, and both the Medical Device Regulation of the EU and the regulation in the US.

The specifics of the IEC 62304 standard will change, they may even take on a completely new shape and structure, but the needs for medical-software development will not change in any hurry. The need to elicit requirements, work efficiently towards delivering a product, meet requirements with the product, and support the product over its entire life cycle will remain constant. The need to maintain trust in what the software does all through its life cycle will be the biggest requirement of all – this should be your North Star or the big neon-colored buzzing sign hanging above your goals on the office walls. That trust is essential for your development work to proceed predictably and achieve the goals set for it, but it will also be the overriding requirement for turning potential users into actual users and allowing them – and your regulators – to rely on your software. This is particularly the case if your software product has an artificial intelligence flavor to it. At a recent FDA event attracting exceptionally comprehensive participation from patients, physicians, industry, and academia the consensus was that transparency is an increasingly important topic to consider in the development of medical devices. Providing all of the development data and all of the information on the development work done will never be possible, nor would it be useful for the end users. However, providing the right information, the right amount of information, and presenting all this information in an actionable form will be crucial throughout the development and subsequent use of any advanced medical device. This is true both within and without your software-development pipeline.

In the coming sections, we will align our radio antennas at the distant twin planets of ISO 13485 and IEC 62304 to figure out an intelligent way of welding the two worlds together. To ensure a good contact is made, just like in the opening scene of that film by Robert Zemeckis from 1997, we will apply both a long history of collected wisdom from regulated software development and a keen ear for sifting through and making sense of recent developments to navigate our path.

Simply put, the goal of this book is to extend the framework set by the ISO 13485 standard on quality management in a way that readily supports the development of software-based medical devices while at the same time adding as little overhead as possible. The aim is to perform all the key activities expected by both the quality management and the software-development standard but to eliminate all unnecessary friction and any duplication of effort between the two domains. To think twice and do once.

The lofty goal here is to take a deep dive into IEC 62304 software lifecycle processes, observe the key processes defined therein, identify any overlap with ISO 13485, and then design a process model that merges the two worlds as safely and efficiently as possible. The underlying premise is that the two standards are fundamentally compatible and have the same goal of ensuring safe and high-performing software products reach the markets. Unfortunately, the two standards have not been written with any real plug-n-play compatibility in mind, which means that forging a unified process meeting both standards requires some careful thought – especially if the intention is to also at the same time use a software-development model that is more agile than gravity-bound.

The good news here is that this sort of coupling is in fact routinely done in the medical-device business. We know that it can be done even if no convenient ready-made models exist to lay out the lines for the alignment. We know that some attempts in the past have been less successful than others. This is where this book steps in, to provide you with a thought-out practical example and an adaptation to tweak to your particular needs.

In writing the discussion in the following sections we have attempted to distill both standards to their cores, expose the underlying requirements, provide practical real-world examples where we thought they would help, and forge a practical, low-friction model for conducting modern software development in a regulated space and under industry-standard quality management. This book will give you the guidelines, sighting angles, and instruments to successfully dock your software-development activities into the rest of your organization's quality management system. Think of the Apollo space program, the lunar landing module, and the command module, but rest assured that no duct tape or underwear containers will be needed here to secure the connection between the modules.

The quality management system itself is described in detail in our previous QMS book (Juuso 2022). The emphasis in this present book will be on extending that quality-management system to also cover software-development activities in a smart way. The base QMS you will be working from here could also be something different. What matters here is that you want to fuse quality management and software-development activities together in an efficient and safe way. The goal is to focus most on what truly matters, to achieve everything you need to while not having to repeat yourself or spread your resources thin.

The following sections of this book will introduce you to the ISO 13485 standard on quality management, the IEC 62304 standard on software lifecycle processes, and a way of adopting both for agile software development in a regulated industry. The discussion is infused with our own experience in the field and the emerging trends that shape the future of both standards.

Our aim is to write the description within these covers in a way that is easily approachable to all interested audiences from customers and management to individual developers, programmers, and suppliers. We hope that the discussion and examples provided in the text will highlight synergies, illuminate possible issues, and allow for the informed design of unified processes across all of the involved activities. The discussion herein is based on our decades of experience in the design, inspection, certification, and everyday running of software-development activities in both industry and academia.

Acknowledgments

After the incredibly warm and encouraging reception to our first book on ISO 13485 quality management for the medical-device industry (Juuso 2022), it was with profound gratitude and an unquenchable thirst for knowledge that we embarked on this second book project. This time we dove into understanding, and making sense of, the required second layer of the puzzle for the development of software-based medical devices: the IEC 62304 standard for software-development life cycles.

We wish to thank all of the people from around the world who have provided their thoughts and reviews of the first book. All of your feedback is greatly appreciated and has, in part, informed the writing of this second book.

This time, too, the people who took time from their busy schedules and provided us with invaluable feedback on the evolving manuscript all deserve our gratitude. Jussi Ala-Kurikka, Tuomas Granlund, and Päivi Turta all gave us in-depth notes on our text and made our scribblings undeniably better. Sandra Liede and Nina Vartiainen, too, conducted a welcome sanity check of the key lessons we wished to depart here. The comments you all gave sharpened our focus, improved our output and added insight we would have perhaps otherwise neglected to include. In the same vein, we also wish to thank our respective organizations, Kasve Consulting and SGS Fimko, for their supportive attitude toward our work here.

Finally, we acknowledge with immeasurable gratitude the understanding and love of our families. Writing this book has not taken place without some weekends and nights spent on Zoom calls and in front of the computer. All of this has, regrettably, been time stolen away from our families, but perhaps the invigorating effects of the collaboration have carried over to family life also. We also take pleasure in the fact that some of the hundreds of hours of conversation have taken place over pints of beer or even in the glow of the occasional fireplace transmitted over a Zoom line. As a quest to consolidate past experiences and learn more about the topics involved this has been an outstanding experience for us. For that, we make no apologies but promise to act better in the future.

As a result of this joint book project, we have both learned a great deal, realized we knew quite a bit beforehand, and come out with a wealth of considered ideas on how to best combine the many standards involved in developing medical software. We hope that we have been able to write this book in a way that passes on some of those ideas and gives rise to new, leaner, and to-the-point solutions at each of your organizations.

About the Authors

Ilkka Juuso has 20-plus years of experience working on multidisciplinary R&D projects in both industry and academia. He is one of the founders of the medical-device startup Cerenion, the Principal Regulatory Engineer at the medical-device quality consultancy Kasve, and a post-doctoral researcher at the University of Oulu in Finland.

His main interests are international regulatory affairs, standardization, and healthcare business development. He has successfully led the development of an ISO 13485, ISO 14971, and IEC 62304 compliant quality management system (QMS) from the ground up, its subsequent day-to-day operation, and certification by a notified body. He has had a key role in the launch of a CE-marked Class IIb medical device based on artificial intelligence.

He has repeatedly served as a committee member and the head of the Finnish national delegation in key committees of the International Organization for Standardization (ISO). He is a member of the management team for CEN TC 251 on health informatics, and secretary for its WGII on technology and applications. He is an active member of the Finnish Standards Association (SFS) and the industry forum Healthtech Finland. He is also the author of the book *Developing an ISO 13485-Certified Quality Management System* (Routledge 2022).

Ilpo Pöyhönen has 30-plus years of experience working on medical-device research, development, testing, and safety and performance evaluation including in the context of an accredited certification body. During that time, he has performed approximately 200 software evaluations according to IEC 60601-1-4, IEC 60601-1 Clause 14, IEC 62304, and IEC 82304-1. His particular areas of interest in this work have been the role of programmable database systems and the development of test equipment for diverse needs. The work has taken him across the globe and even to the edge of space.

Today his main interests are international regulatory affairs, standardization, and the intelligent control of medical-device design processes to continuously meet the requirements imposed by, for example, cyber security, usability engineering, risk management, and agile development models. The use of emerging technologies, such as artificial intelligence, also holds special appeal to him.

He has been active in research initiatives that have, for example, examined the software-development documentation required in a regulated environment, the impact of risk management, the performance of risk analysis itself as part of the software development life-cycle, and the reliability factors involved in the supply of complex software systems.

He is a long-time committee member of SFS/SR301 on healthcare IT and a sought-after lecturer on topics such as medical-device software, risk management, usability, mHealth apps, and cloud services in the context of medical devices.

Chapter 1

WHAT TO KNOW BEFORE GETTING STARTED

The journey's start is looming just beyond this section, but before we embark, there are a few key things worth pointing out. This section will list some of the most valued lessons we have learned over the years, and hopefully provide you with welcome insight on the path to designing the most streamlined and thought-out development pipeline for medical software.

In the army, new recruits are first broken down to clam chowder and then rebuilt in the image of super-soldiers once everyone is on the same page. Something similar happens with the gluten in flour when making noodles. The purpose of this section is not to break you down or brainwash you into adopting standards and our views on these two particular standards, but to arm you with the arguments you may need to convince yourself and others on the merits of combining IEC 62304 and ISO 13485 in the development of medical-grade software products.

In addition to the lessons set out here you will find a host of other fundamental lessons mostly drawn from implementing and working with an ISO 13485 quality-management system in our earlier book *Developing an ISO 13485-Certified Quality Management System* (Juuso 2022). The discussion in that book will be particularly useful to you if you are now figuring out the relationship between quality management and software development, or even writing the standard operating procedures for software product realization yourself. If you are interested in just understanding software development itself, without consideration of the surrounding quality management context, the discussion in this present book will likely suffice.

This book will concentrate on the IEC 62304 standard that acts as the often-quoted backbone to the development and maintenance of software in the context of medical devices. The discussion in the following sections will distill the essential concepts, requirements, and solutions of the standard and fit these within the wider context set by the ISO 13485 standard on quality management. Both standards are de facto expectations in medical devices for much of the world, and countless organizations need to adopt and adapt the two standards into their own operations. The way to combine the standards into a lean and frictionless setup for your organization is the goal here. Step one is to understand both standards, and the first step in that is to dispel a few easy misunderstandings. This section will do just that to ensure we are on an even keel and get off to a good start.

1.1 There Is No One Correct Answer to Anything

This may sound like it pulls the rug from under this whole book, but it really doesn't. Understanding this one simple lesson will let you begin to see the matrix of intentions and goals behind the standards and all regulations and to begin crafting your operations on top of that network. Don't get stuck looking for that one needle, there is more than one in the haystack. In other words, the sweet spot is wider than just one hole to hit on a golf course, as Figure 1.1 illustrates.

Not all roads lead to Rome, but more than one does. This all demonstrates that you can adopt a standard in more ways than one. To any question you find in the standard you can bet there is more than a single equally acceptable answer. Use this realization to save you from looking for that nonexistent single needle in a haystack, and to give you a sense of calm in knowing your own pockets when constructing your processes according to the standards. Don't use it as a call to only do as you will, though.

After you have answered a few such questions you will have started to make design choices about your overall approach that probably affect your later answers as much as anything you then read in the standards. You must understand the standards and meet the requirements given in them, but how exactly you go about doing this, which answers out of the pool of possible

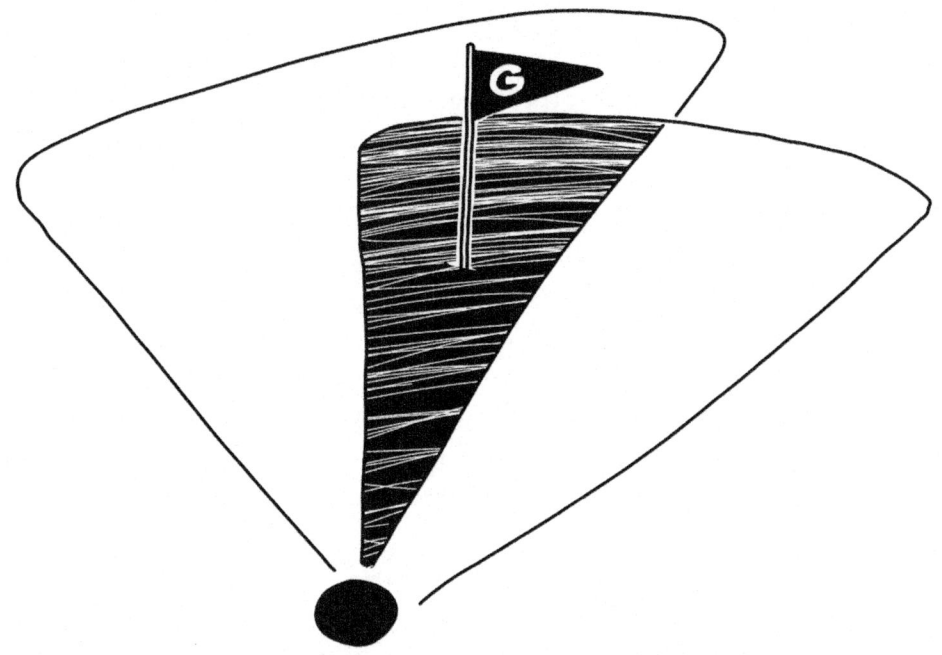

Figure 1.1 The sweet spot is wider than just a single hole to hit.

answers you choose for the implementation of your processes depends on the context given by your specific environment. The one universal characteristic of any resulting good development model is that it ensures you meet all the requirements and create the required deliverables consistently, and not only as an afterthought to how you actually want to work every day. Otherwise, you will not be using your resources, whatever they may be, in any optimal way, and you won't be well-equipped to deal with any future changes to the environment. Doing something extra not required by the standards is not a horrible state of affairs if your resources allow it and your peace of mind requires it, but doing one thing and then duct-taping something else together based on that is not lean, it's not smart, and it probably isn't safe either.

Trying to create an ideal showcase, a perfect development process with perfect documents to show your auditors, is not a noble goal either – particularly if your showcase does not correspond with the back office for any reason. If the showcase you have created does not lead to a lean, efficient, and safe software development process, or if it doesn't jibe with your daily operations, it is just a plain bad idea. Perfection only lasts in fairy tales – the grind of the everyday will call that "happily ever after" into serious question sooner or later if you're not economical and strategic in your designs.

Another occasional sin here is to write abstract, lofty processes thinking that you don't want to accidentally require doing more than the standards call for and thus should stick to the language of the standards as closely as possible. This is a fine theory, but one which may fail to translate the requirements onto a practical level. As a result, you may end up having to reinvent the wheel every time you then invoke a process, which will surely waste resources and might even have you coming up with some oddly different wheels over time. In practice, you'll want to find a good, concrete level of defining processes so that they are actionable, safe, and to the point. Knowing why the processes are run is the key to finding this level. Most everyone knows the story of Icarus, the winged boy from Greek mythology who flew too close to the sun, but not everyone remembers that he was actually instructed to fly high enough to not dip into the sea and low enough not to melt his artificial wings in the sun. There is both a ceiling and a floor for flying in that story, and the same is true for constructing your QMS. This is also a shortcoming with some of the lesser-known standards out there: some are much too theoretical for practical application and others much too micromanaging to negotiate in the real world. The good news here is that the standards discussed in this book are neither.

This book highlights the design choices you will be faced with and provides you with suggestions and possible answers based on what has worked in the past, but the answers you ultimately choose will be up to you. Lesson one, therefore, is that just because something is standard it isn't necessarily uniform, or all cut from the same mold. That famous Apple advert called *1984* was not aimed against standards but against unquestioned monotony. Standards are our best chance of ensuring pieces fit together and can be both inspected and understood from the outside in, but the way those pieces are constructed can be as varied as life itself. In fact, DNA too can be thought of as a standard, but the life encoded by that code could not be thought of as monotonous.

So, how do you recognize the wrong answers within the pool of the many possible answers? How do you split the wheat from the chaff, in other words? Any answer that lets you dismiss some requirement that you don't quite understand is probably going to be a fake answer akin to burying a timebomb under the foundations of your house or leaving a leaking water pipe alone in an area prone to sinkholes. On the other hand, an answer that lets you be a little lazy may not be a horrible starting point if you understand the requirements and are confident that your solution achieves what is needed efficiently, reliably, and in harmony with the rest of your operations.

The standards call for a wealth of aspects to be considered throughout your operations, but never will they insist on a separate, single record being created in a vacuum and for the sake of itself. You must know how the standard relates to your operations, but the standards are not your project plans. The IEC 62304 takes this to one extreme by stating that you should map your processes to the requirements of the standard, but you don't have to adopt the processes as such.

In general, one of the best excuses for developing software in the first place is that it lets you instruct something once and have it performed for you over and over again without raising much more than a fingertip if even that. Just like in developing software where copy-pasting algorithms all over your code, working on too many things at once, or creating new separate functions for each new identified requirement, the same is true for looking to comply with the processes and deliverables described in a standard. Think of any standard as a new comb to use in sifting through your own operations and identifying the points of common interest to shape your answers. Some standards you can use as inspiration and others you must follow closely enough to reach compliance. No standard is a project plan for a business, though. At best a standard could be thought of as the inflatable raft that you breathe life into to cross the waters to where you want to go.

Really the choice between the possible implementations answering to any particular requirement from any standard comes down to you feeling confident you know all the touchpoints around the issue, and that it does not feel like you are brushing anything under the rug. Only realize that the person coming to look under the rug at the end will be your customer, your certification body, or your regulatory authorities – and that lives may be at stake.

1.2 Why Do We Need a Standard at All?

> *If there is no one correct answer, then why do we need a standard at all? We've heard this asked every once in a while, and we've even heard senior consultants argue a version of this occasionally. Standards, though, allow us to plot a course, as Figure 1.2 illustrates.*

The premise is that manufacturers will do the right thing anyway, and a micromanaging standard just gets in the way of that and detracts from making a better product. To a certain extent this is true, splitting attention and

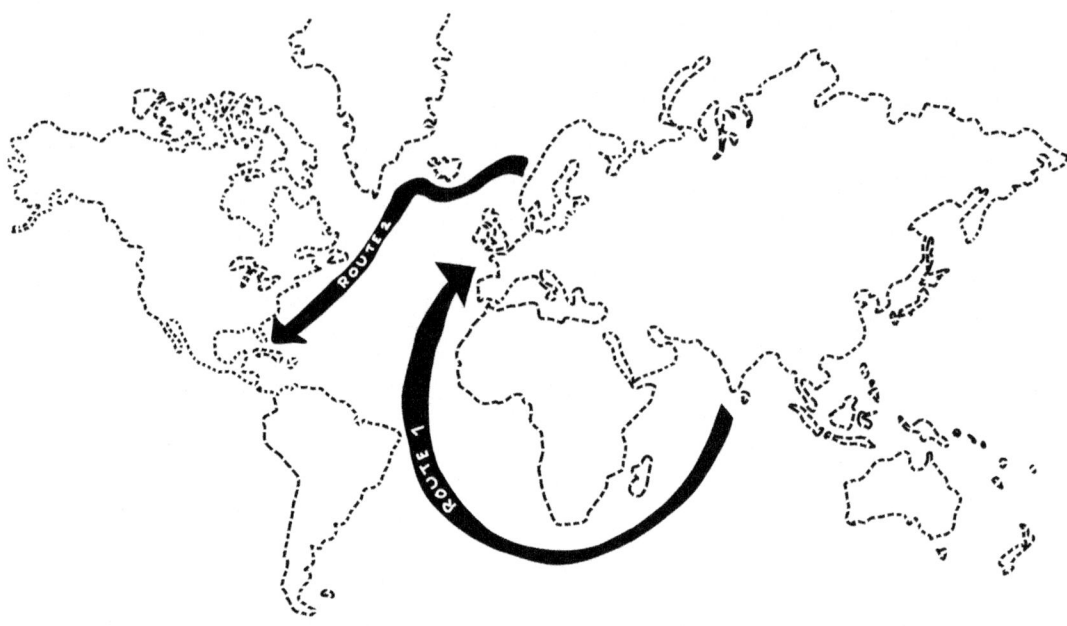

Figure 1.2 Standards are like maps to the otherwise uncharted world.

distracting anyone from doing their work is certainly to be frowned upon – or squished mercilessly like an unwanted bug that creeps into your code. Doing more different things will, without fail, mean getting less of each thing. Having to do something some way that feels stupid, silly, or just cumbersome is never a goal.

The problem is that "anyway". It is arrogant to the point of extreme to suggest that, if left to our own devices, every manufacturer would consider all the same poignant points, solve any design choices in a compatible way, and be able to explain what they have done using language and concepts readily accepted by the users, partners, and regulators out there. All of this is why standards are such practical tools: not only will they help ensure you have covered all of your bases, but that everyone else will speak the same language and have the same fundamental expectations. From a conformity assessment point of view, standards make it possible to formulate requirements in a universal way that then lends itself to a more straightforward assessment. In other words, we need standards to meet common expectations and to achieve a good level of interoperability everywhere where it matters.

The use of IEC 62304 is optional, but in practice, you should be prepared for long discussions and endless surprises if you opt out. Use of the standard is expected for any software touching on or taking the form of a medical device. IEC 62304 is, therefore, all but required. As for some other key standards, the same arguments apply. IEC 62304 bootstraps a few additional standards into the mix. It strongly implicates the ISO 13485 standard for the quality management of your overall operations, even if it stops short of requiring it (see Section 4.1). The present version of IEC 62304 takes an even stronger stance on risk management: it casts your feet in a vat of concrete and throws you into the pool of ISO 14971 risk management whether you feel like it or not (see Section 4.2). Think of the opening scene of *Billy Bathgate*, the new-found clarity on life Bruce Willis gains with his new footwear and embrace your new standardization overlords as friends. Or be prepared to argue your case for individuality like Mel Gibson did as William Wallace in *Braveheart*.

All joking aside, standards may seem like something extra to do, but they should not be that. Standards should help you inspect your processes, prioritize between aspects of the processes, assign resources, and improve on the processes to better achieve your goals – all the while making them instantly comprehensible to your stakeholders. There are tens of thousands of standards out there, not all of which you should ever adopt, but the select few featured in this book really are industry standards that you should give serious consideration to. Here our focus is primarily on the two standards, IEC 62304 and ISO 13485, which we see forming the foundations for much of the medical-software development work today. We will also touch on a few others occasionally but treat these references as beyond the scope of our discussion here.

Following standards is not mandatory, but it may be in your interest to do so. Following a standard will increase your likelihood of considering all the poignant angles and developing a solution that is compatible with other stakeholders. At a recent industry event, a representative of a competent authority went as far as to imply that standards can be met selectively. This is certainly true if you don't plan on reaching conformity with the standard, but instead only apply it as a source of inspiration for your operations. If conformity is your goal, however, pay careful attention to how each standard is allowed to be sampled, what it itself and your applicable regulations say on the matter, and what this may mean for your other related operations. If you are in the EU, of special interest here are the harmonized versions of each standard and the special Z-annexes made for them, if these are available. The Z-annexes of the standards will be valuable tools for you as you look to meet regulatory requirements in the EU by utilizing the standards. Also know that another concept, the so-called "state of the art" (see Juuso

8 ■ *Medical-Grade Software Development*

2022), will be your friend if that harmonization process is lagging behind and you want to use a newer, but already commonly accepted version of a standard instead of the archaic version which may still be the latest version officially harmonized with the EU regulations. Don't do so on a whim though, but rather after careful thought as the European system is intended to use harmonized standards.

1.3 What Is Medical-Grade Software?

Software used to meet a medical need is naturally subject to stricter requirements than your run-of-the-mill Angry Birds app, but how do you know when a piece of software should be considered medical in nature? There are multiple dimensions to defining the concept, as Figure 1.3 illustrates.

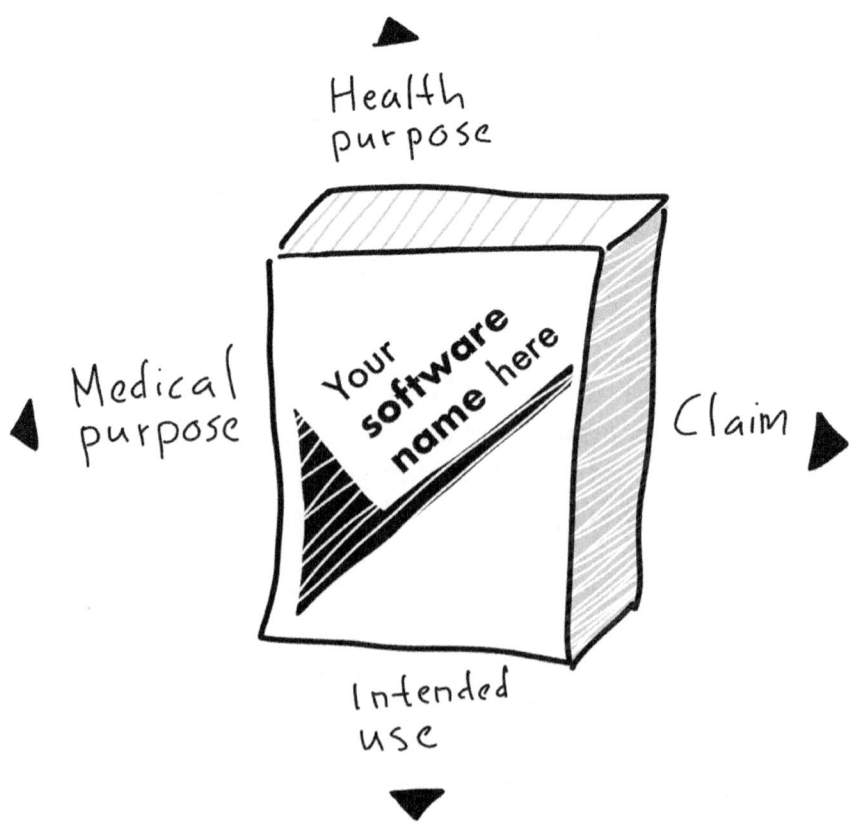

Figure 1.3 The definition of "medical software" is a slippery concept to nail down.

This is a big question, one almost on par with, "what is the meaning of life"? The line of demarcation between medical software and other software is the equivalent of a modern-day battle line separating two belligerent camps. It is the entrenched line between what is a medical device subject to strict regulation and what is a mere app or device for some wellbeing application. It is the line between what is expected in the regulated domain of medical devices and what just about anyone can publish in an app store by themselves.

Even the above is a gross simplification. The fact of the matter is that what qualifies as medical software is not governed by standards, but by regulations instead. Those regulations will change from one geographic jurisdiction to the next and will even differ based on how the manufacturer has worded their intended use and marketing message for the software. In Europe, the Medical Device Coordination Group (MDCG) acting under the European Commission has put out helpful guidance on determining what software qualifies as a medical device (e.g., MDCG 2019-11) leaving out, for example, electronic health-record systems that store and display the entered data without layers of machine-generated analytics in between. Similarly, the Medical Device Regulation (MDR) itself, and particularly its Article 2, affects what software should be considered medical and what not. In the US, the FDA has made a wealth of guidance and thinking available on these matters via its website.

IEC 62304 takes a pragmatic stance on the matter – no doubt sheltered by the years that elapsed between its minting and the present debate. The standard defines medical-device software as something developed for the purpose of either being incorporated into a medical device or acting as one itself. The former could be called Software in a Medical Device (SiMD) and the latter Software as a Medical Device (SaMD) in present parlance. It has occasionally been argued that IEC 62304 doesn't readily apply to standalone software, and this is reflected in some of the goals of the standard (e.g., the suggestion of synchronizing risk-management activities over both the software and its host device or performing additional validation activities on top of the work the standard instructs), but this does not have to be the case. The standard today is routinely used to assess different types of software from firmware to software as a medical device (SaMD). It is noteworthy that even an assessment of a hardware device based on IEC 60601-1 requires the use of IEC 62304 for any embedded software components to reach its goal (see Clause 14 of IEC 60601-1).

Another part of the debate is whether medical devices are included in the definition of health devices. Most experts would agree that these are two opposite ends of the same swimming pool, but they would be hesitant to say that wading the shallow end would set you up for surviving the dive boards at the other end. Similarly, you might look silly and be chronically out of breath if you geared up for the deep end and spent your time at the shallow end. Both medical software and health software may have some aspect approaching a medical purpose or saying something about an aspect of health but treating them the same is problematic to say the least. The attempt to do so appears to also be one of the major factors behind the delay of the new version of IEC 62304.

Following the same hardcore model of software development expected of the highest-risk class of medical software when developing a low-risk health app is not the best use of your resources – and requiring manufacturers to do so will stop many development projects dead in their tracks. The loss will not just be monetary, but a tangible lack in the availability of solutions and the benefits they could bring to the health of every one of us. Viewing your software-development operations through the lenses provided by IEC 62304, ISO 13485, and even ISO 14971 risk management will, however, show you where the paths diverge and what you need to do when moving between the ends of the swimming pool.

IEC 62304 makes a distinction between "software" and "medical device software" in that the latter is software – or a software system, to be exact – intended to be used in a medical device or as a medical device (Clause 3.12). This is compatible with what ISO 13485 considers a medical device (its Clause 3.11). Not all software, even inside a medical device, has to be medical-device software, though. This all depends on what your product is, how you design the software architecture, and how you implement the features corresponding to your intended use. The software that implements your intended use definitely is that, but some supporting software may not be. This much is also hinted at by the MDCG 2019-11 guidance.

1.4 Software Is Classified by the Safety Concerns Associated with It

Tom Hanks had to put a leash on Hooch in that film to prevent wanton havoc to his apartment, and any time you give Schwarzenegger a big gun you can bet there will be gratuitous destruction. Both Hooch and Arnie should thus come with a warning, an inflatable sumo suit, or just some instructions for careful use. The software safety classification instructed by IEC 62304 is that added caution, as illustrated in Figure 1.4.

In the previous lesson we started to lift the lid of the Pandora's box of what software goes into your medical device. The standard offers a three-level safety classification you apply to the software as a whole, but also to the various subsystems underneath. Think of this as the different colored high-visibility vests given to your software components based on what level of risks or safety concerns are attached to them. It's not a license to goof off or a radio beacon to stay clear, but it is a tool to help you allocate the necessary attention and resources to where they are needed.

The ABC classification, from low risk (A) to high risk (C), is a fundamental feature of the IEC 62304 standard that changes how the requirements in Clauses 5 through 9 of the standard apply to your software. In other words, your understanding of at least 58% of the contents of the standard may be affected by the safety profile of your software. In practice the figure here is lower as many requirements are shared between the classes, but occasionally the differences are significant.

The classification is not a quality stamp on your software, class A software is not better than class C, but it does allow you to focus more of your resources on the development of class C software where risks and potential harms may be greater. The classification is addressed in detail in Section 4.3

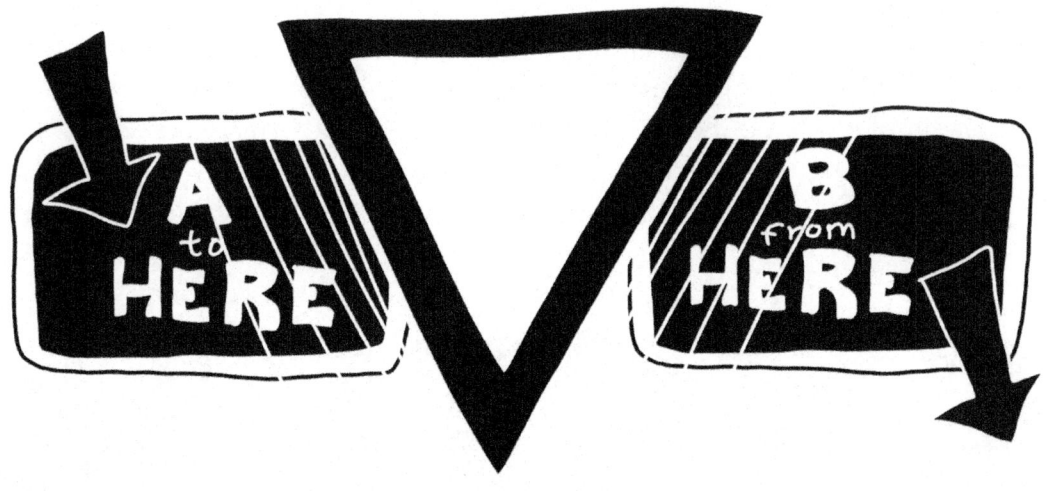

Figure 1.4   It's not quite as easy as in that nursery rhyme, but knowing your ABCs is required.

of this book, but it is worth being aware of this undercurrent right from the start.

Note also that the FDA has its own old three-class Level of Concern (LoC) rating and a new two-class Documentation Level rating to categorize the risk of software as a medical device. Neither categorization is directly related to the ABC classification of the standard, although both are compatible. Instead, if your market area covers the US, you must follow the guidance given by the FDA, answer a set of specific questions listed in the guidance, and arrive at a LoC rating for your software. Also noteworthy is that not only is the LoC categorization different from the ABC classification given by the IEC 62304, but it is also separate from any medical-device classification (e.g., Class I–III). The good news is that all three are compatible, even if not identical as mechanisms.

1.5 Software Is Made of Units, Items, and Systems

Here we start to lower our diving bubble, like Jacques Cousteau or Jules Verne might have done, into the IEC 62304 standard by looking at how it defines the construction of software. On the way down to the core of the constellation of a software system we'll encounter items and units, as illustrated by Figure 1.5.

The standard takes an engineer's approach to understanding concepts by dissecting them into smaller and smaller elements until you feel like you can't continue anymore. The French philosopher René Descartes (1596–1650) would certainly be proud of this drive to reduce complex objects into collections of simpler interconnected objects. Intuitively, too, this makes sense for understanding the inner workings of software. On a first dive, though, the sea of interlocking terms may seem bewildering. For this reason, this book

Figure 1.5 System begets an item, an item begets a unit.

will introduce you to the essential terms and concepts as we progress, not all at once, but when you first need them.

The first set of terms to know are software unit, software item, and software system. These are all terms used when dissecting software and will be covered in detail as part of the terms and definitions in Section 3.2.5. For now, it suffices to know that what we typically discuss as a piece of software is actually a software system in the standard's parlance. This is the entirety of the thing you are looking at as your software, whether that is the whole of your medical device (e.g., software as a medical device) or just software running inside (e.g., software in a medical device). This can then be dissected into software items on a level of granularity that best suits the conversation or the design and development (D&D) of the piece of software. There can be several nested levels of software items, each more refined in granularity than the previous. Ultimately, when an item can't be broken down into any more composite items it may be called a software unit. In other words, units build up items, items build up software systems.

One straightforward way of thinking about the granularity of software items is that a software unit corresponds to a class in an object-oriented programming language, a software item to a library (perhaps containing frontend or backend components), and the software system would then be the whole software itself. This is, however, not the only way to set up granularity. It might be cheating to say that your software can't be dissected into any items, to say that in your unique case, your software system is your software unit, but many other definitions of depth may be acceptable. If your software comprises one long legacy algorithm with a completely flat structure to it – without even a single GOTO instruction inside it – this might really be the case, but for most modern software that is not going to be a defensible position. Creating superfluous structure in your hierarchy will not be practical, but at the same time it may also be questionable if you would actually save any effort by squishing your hierarchy unnecessarily here.

The slightly mad thing in the scheme is that all these various granularity levels can be called software items. In other words, a software system can be thought of as a software item, a group of nested software items can be thought of as a software item, and even a software unit can be thought of as a software item. The term is thus used quite liberally to refer to some identifiable software object. Don't worry, this only sounds mad but is quite manageable in practice. In a way, this is like the fourth *Matrix* film where we have already had three installments of the story but are now told that it was all a game, led to a video game, and can all be discussed on video via talking about the game. It's nuts, but it's also certifiable and we can live with it all.

The above amorphism is, to some extent, a relic of the evolution of the standard. In the IEEE 610.12:1990 standard, which forms a part of the ancestry of the IEC 62304 standard, software architecture is described using the terms system, component, unit, and module. The system there is said to consist of parts which may be hardware or software, and which may be further subdivided into other parts. The terms used for the parts are, for example, module, component, and unit – all of which are noted to occasionally be used interchangeably and in varying order. IEC 62304 thus has a broader understanding of what a system is (e.g., it can include not just software and hardware, but even the user themselves), a more focused concept of a software system (just software), and the same fluidity towards what parts a software actually consists of (all of which are now called software items). The most striking difference between parent and child here is that IEC 62304 continues to think of a software unit as a logical block that is not subdivided, but it does not directly speak of the testability of this block – it does speak of unit verification later (see Section 5.5). Understanding this legacy here is above and beyond what is expected to develop software today, but it does demonstrate that the relative amorphism here was not accidentally introduced by IEC 62304.

1.6 Software Development Has a Life Cycle

> *The* Fast & Furious *saga may keep on ticking and racking up those version numbers, but it doesn't really evolve in any discernible way. The frequent updates are a sign of a well-oiled machine that knows its intended use and formula. Similarly, developing software must be followed with the appropriate steps in its later life cycle, as depicted in Figure 1.6, to meet all the expectations of the users, customers, and regulators.*

IEC 62304 prides itself on providing a model for the software-development life cycle. But what is a life cycle model anyway? Is software supposed to be conceived, be born, grow up, produce offspring, retire, and ultimately fade away? Will the software then be reincarnated as something else? Are the medical-device labels "single-use" and "use by date" then the most

What to Know before Getting Started ■ 15

Figure 1.6 Not to put too fine a point on it all, but a life cycle is just that.

vicious marks of all to be placed on such living software? Until we are faced with autonomous, self-aware software systems we won't have to get that philosophical. Software for now is a thing, it doesn't have life goals to meet, it doesn't have an Aristotelean drive to do anything by itself. Software is not alive until HAL 9000 greets us as Dave. The life cycle is just a way of modeling the path from conception to a releasable piece of software, and the steps along that path.

In software development, as in life, the usual stages of existence can, however, be characterized as a path from gestation to launch and then constant improvement during use until decommissioning. Both the IEC 62304 and the ISO 13485 standards intend to ensure you know what you are launching out into the world, and then make sure you don't forget about it once it is out there. You are even expected to make arrangements for the funeral service before the clock is started.

The IEC 62304 defines the term software development life cycle model, but it doesn't define a term for the entire existence of software. Clause 3.24 defines the arc of a development life cycle from requirements to release. In this conceptual structure, you model the following:

- Identify the processes (incl. activities and tasks) involved
- Describe the sequence of the identified activities and tasks, and any dependencies between them
- Identify the objectives of the activities and tasks, i.e., the desired deliverables (i.e., the required result, output, or documentation of an activity as per Clause 3.6) and set the milestones for verification of their completeness

In the definition above it is striking that the later maintenance of software is not included in the life cycle and no separate entry for the software-maintenance life cycle is made in the standard. Rest assured that requirements for maintenance are placed on software though, and these are exacted through, for example, Clause 6 of the standard on maintenance (see Section 6) and the ISO 13485 standard. In no case is it enough to just build the software and let it float off to the sunset – unless we are talking about a piece of non-medical software like the *E.T.* Atari game, in which case no users are involved in the first place. Software deployed directly to a landfill notwithstanding, the life cycle is a broader concept than just development.

The ideal development processes may have a life cycle model, but increasingly the development phase is followed by an equally important maintenance phase. Software is no longer just built and reused, but instead, it evolves throughout its true life cycle. This is particularly true in the case of any methods resembling artificial intelligence built into your software. Here the regulators will be looking at the long-term life cycle of your medical device.

The seeming short-sightedness of the life cycle as defined in the IEC 62304 standard is somewhat baffling, but the afterlife awaiting the software after launch is an increasingly weighty topic today. In fact, the future we envision for artificial intelligence in software, and other medical devices that may one day self-evolve out there on the field once launched, is placing some pressure on the concept of the life cycle to evolve. The FDA appears to be the frontrunner in addressing this new post-launch evolution with initiatives such as its Pre-Cert program set to enable some evolution to occur within predetermined limits once a device is already in use.

1.7 Everything Is a Process, and Those Consist of Activities and Tasks

> *A process is the ultimate Lego block in any standard. The process gives structure to how you move from an input to an output in a way that achieves both some value-adding work in between and compatibility with the other Lego blocks of your system. A process, then, can be thought of as consisting of activities, and these consisting of tasks as Figure 1.7 illustrates.*

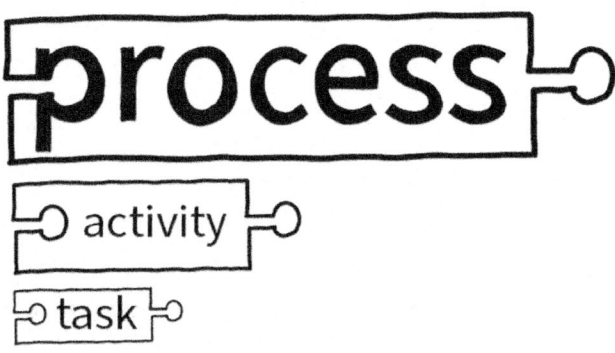

Figure 1.7 Everything is a process.

In another exercise in terminological reductionism, the IEC 62304 standard sees all operations as fundamentally consisting of processes to run. These are in effect the chunky large nautical shipping containers for your work. IEC 62304 is not alone in taking this view of processes as Lego building blocks, the ISO 13485 standard too considers value-adding work from an input toward a desired output to be a process. Processes are then stacked to fill your cargo ship, or in this case, instruct your quality management system (QMS) and your software development pipeline. The IEC 62304 standard goes on to instruct that all processes consist of activities, and these in turn consist of tasks. The terms process, activity, and task are addressed in more detail in Section 3.2, but for now, it's enough to be on the lookout for this triplet of terms and know that they too form a Russian doll of sorts.

There is one small exception to the above rule of thumb: instead of a defined process, you might occasionally have documented arrangements to instruct the handling of some activity. ISO 13485 mostly speaks of defined processes and procedures, but on a few select occasions it shies away from requiring a process and instead accepts a set of documented arrangements. These occasions are contamination control (Clause 6.4.2) and communication with customers (Clause 7.2.3). The standard also speaks of the importance of adherence to documented arrangements in product realization as part of D&D reviews (Clause 7.3.5), D&D verification (Clause 7.3.6), D&D validation (Clause 7.3.6), and the monitoring and measurement of the product (Clause 8.2.6) – and it requires these arrangements to be satisfactorily complete before product release and service delivery take place (Clause 8.2.6). Such arrangements are also to be a part of your internal auditing activities (Clause 8.2.4). In practice, the difference between a process and an arrangement may or may not be significant for your operations.

1.8 IEC 62304 Does Not Insist on a Specific Process Structure

As illustrated in Figure 1.8, processes slot together in some meaningful way as instructed by your QMS and the IEC 62304 standard. That creates the system in a quality-management system or the pipeline for your software development activities. Here you should take your ISO 13485 QMS as the backbone and ensure your instructed processes also meet the requirements of IEC 62304.

The introduction of the standard (see Section 3.2) makes it clear that the standard assumes all work to take place within both a quality management system and a risk management system (see Sections 4.1 and 4.2, respectively). Within this context, it then defines a fundamental four-process framework for the safe design and maintenance of medical-device software. The four processes it presents are:

- Development (see Section 5)
- Maintenance (see Section 6)
- Configuration management (see Section 8)
- Problem resolution (see Section 9)

The IEC 62304 standard does not force this specific model of processes on its users but instead expects the user to map the requirements it does make on any of the processes (incl. activities and tasks) to the life-cycle model they have adopted. This allows for great optimization in fitting the standard and the real world together, but it will also be a major topic of discussion when comparing the operation of any two manufacturers – not to mention a major undercurrent of any conformity assessment involving the standard.

Figure 1.8 The pieces of the puzzle are up to you.

1.9 IEC 62304 Does Not Impose a Documentation Set

The standard is secure enough in its requirements that it doesn't much mind if you meet those requirements using a novel set of documents. As long as you can map the requirements of the standard to your documents and processes you may christen them however you like. The naming convention of your document is thus up to you, as Figure 1.9 illustrates.

The IEC 62304 standard requires that tasks are documented, but the shape of this documentation is left to the user of the standard. No names, formats, or explicit content is specified for the documentation to be produced in theory. This is a significant design feature of the standard and one which you may miss if you skip learning about the introduction of the standard (see Section 3.2). The standard does discuss a few essential documents, such as the software development plan and the software-maintenance plan, in detail, but it is happy to find the same information in a document by another name or as part of some other documents. Depart from the suggested optional document names at your own peril, though. At the very least you will be the recipient of that famous Clint Eastwood squint from his westerns if you decide that, like the man, your documents need no names.

You should also realize that each review talked about by IEC 62304, or by ISO 13485, must be documented somewhere and thus should be represented

Figure 1.9 Is a document by any other name not still a document?

by a report or other record in your documentation set. Similarly, the regulations applicable to you place distinct expectations on what questions your documentation set should answer, e.g., via Annex II of the EU MDR. How you meet these requirements is up to you, but as time is money – more specifically all time including your auditors' time is your money – it pays to meet the requirements as predictably and straightforwardly as possible. It's not even about money, not really, it's about knowing what it is you are doing, retaining that information also for later use, and keeping the pool of documentation tidy and ready for use over the years to come.

1.10 Applicable Regulations as Part of the Requirements

Any time you see the term "applicable regulations" in a standard you should feel Judge Dredd peeking over your shoulder and see Stallone's muscular jaw protrude into your field of vision to force you to think carefully about what this actually means. Standards are a big part of meeting regulatory requirements the world over, but they can't give you a free pass regarding the laws and regulations – especially not the local laws across the globe. Figure 1.10 illustrates the stack of expectations placed on you.

Figure 1.10 Applicable regulations are a foundational part of your requirements.

Regulations place a great many specific requirements on your work, but the most fundamental one of these is that someone from the outside must be able to trust that you have attempted to create a good piece of software that is useful for someone out there – and should that someone have issues with your software you will address those to the satisfaction of the regulators.

Regulations don't generally add any specific requirements to your software-development operations that you wouldn't already come across in the standards. Safety and efficacy of the produced device is a major concern that cuts through everything, but for the most part, the magnification lens here is on the final product itself and not the process leading to it. Regulations can set some very detailed requirements related to, for example, device classification, device registration, and reporting, but these too do not necessarily concern your core development process. Identifying the relevant regulatory areas of concern may be aided by regulatory guidance given on the type of device you intend to produce or some general safety requirements as are, for example, laid out in the General Safety and Performance Requirements (GSPR) of the EU MDR. Understanding your applicable regulations is something required even by the standards, and an area where understanding your requirements will enable you to fashion the smartest software-development pipeline, or finetune the pipeline we will be discussing in this book. Expect the following to be topics on how regulations affect your work around software products (i.e., your processes):

- Device classification
- Device clearance
- Device registration
- Device lifetime
- Document and record retention
- Vigilance
- Post-market monitoring

In general, IEC 62304 has enjoyed a status as a harmonized standard in the EU (originally harmonized according to EU MDD) and as a recognized consensus standard in the US. The ISO 13485 standard is not only a fully harmonized standard in the EU (both in terms of MDD and MDR) but also presently fast tracked for use in the US by published guidance from the FDA. The other key standards, including ISO 14971 and IEC 82304-1 appear to be progressing along the same path.

In addition to regional authorities, you may also want to keep an eye on the International Medical Device Regulators Forum (IMDRF) and the guidance they publish. We recently had the opportunity to take part in an IMDRF symposium organized by the European Commission and were pleasantly surprised by the practical, common-sense, and no-nonsense attitude that cut through many of the presentations given by regulators from around the world. The takeaway message from the week was that the medical-device world is increasingly harmonized and connected – this means synergies in both the assessment of medical devices and across the surveillance systems built for them around the world.

Marrying regulatory requirements with the development of medical-grade software (i.e., medical-device software) is a topic discussed throughout this book, but especially in Sections 10 through 13.

1.11 Building on the Quagmire That Surrounds IEC 62304

Rome was built on a swamp, but it ruled much of the known world for over a thousand years. St. Petersburg, Russia, was built over a swamp too, and the beginnings of Washington D.C. weren't too spectacular either when the founding fathers got started with it. All this is to say that the IEC 62304 may not be stable for a millennium, it may not be all that we would like it to be today, but it is the code much of software development in a medical context is expected to follow. You can build on it, but you will do well to consider how you build on it. Figure 1.11 illustrates this setup for IEC 62304.

Nothing about the revision of IEC 62304 has been speedy, so let's not call it quicksand, but the fact of the matter is that it is an aging standard last given a partial facelift in 2015. It is nonetheless a backbone for developing medical software and vast numbers of developers, regulators, and certification bodies rely on it.

IEC 62304 is a bit like the legendary film actor Morgan Freeman: you know that he is getting on, but every time you hear his deep soothing voice you feel like a baby and take it as the voice of God. The standard has the same function. It is not perfect, it is not even up to date, but it is the best thing we have for understanding how software development is required to take place – and how it is currently taking place – in medical-software companies. To discard the standard would be to leave a great many developers

Figure 1.11 The IEC 62304 has reached a point of stable uncertainty.

without a compass, and to miss out on providing them all with a migration path to something more modern. The future of the standard looks uncertain and may be subject to even drastic moves, but all the fundamental needs are described in the present version of the standard.

The ISO 13485 standard by comparison is a de facto global standard in the wider context of operating in the medical devices space. The development of medical-grade software should fit within this wider context. This standard devotes much of its column space to product realization, which in the form of Clause 7 of that standard takes up 34% of the entire standard. This standard is a global superhighway between different jurisdictions and stakeholders, and while not all those intersections have yet been christened, it appears to be our best hope for a truly international approach to medical devices.

Consequently, our suggestion is to start with ISO 13485 to build your overall approach to quality management and then to make sure IEC 62304 requirements are met within that grander context. This way you at the very least know how your operations match with the expectations of both standards, and how the two are aligned in terms of software development. Even if the wildest ideas put out there come about, and IEC 62304 is indeed fused with ISO 13485 in some way, you will be in a good place to revise your processes using ISO 13485 as your backbone. This book will help you to see the linkages and possible disconnects between the standards.

1.12 Retaining Trust at Each Step of the Way Is Perhaps the Most Critical of Tasks

In movies anytime you hear the question, "Do you trust me?" you can bet that something risky is about to happen. This happens in Blade Runner, National Treasure, Aladdin, *and countless other films where a momentary act of blind faith will end up saving the day. Blind faith does not match well with the development, adoption, and use of medical-grade software. Instead, evidence is always needed in one form or another as Figure 1.12 illustrates.*

In case you don't have the budget for getting Morgan Freeman or James Earl Jones to sweet-talk both you and your customers into living with your software, being able to demonstrably meet expectations will be a central topic of everything you do. This applies to you trusting the rest of your development team, your regulators maintaining trust in you, and your customers knowing that they are getting what they want all throughout purchasing and using your product.

Figure 1.12 Steps to trust may be different in height, but the action is always the same.

At a recent FDA event on the adoption of artificial intelligence in the context of healthcare, the special topic of building in transparency came up as a particularly poignant goal. It will never be possible to provide all the development data, documentation, and source code of your software to all its potential users, have them take the required time and apply the required acumen to assess the whole, and after that make the decision to purchase and use your software. This would be a pipedream of the highest order, and while easily required, never actually implementable in real life. Instead, you have to be able to obtain and provide enough information on what is happening inside the box for the person outside the box to stay confident they know all they want to know and that it is all taking place as they want.

Transparency is needed for trust in and adoption of the results of the box, but we also need that information to be made available in an actionable form. Overwhelming the user, any user, with the quantity or complexity of information is not helping anyone. We need the right information, the right amount of information, and that information to be presented in a readily actionable form.

The need for transparency and appropriately timed information cuts through the whole software-development cycle. The classic waterfall model of software development meets the need of obtaining reviewed and approved information about what has been done, which is ultimately what everyone wants: to be able to trust the thing you are handed. It does not, however, provide for a good real-time view into the development process of software or account for accumulated knowledge and increased certainty over time. Partly for this reason, other more agile methodologies are often used in software development today and can in fact provide for an even greater level of transparency during development. The threaded and iterative nature of agile development will cause some head-scratching, particularly during the development of the processes themselves, but when done right, agile can be a win from all points of view.

The argument can be made that "retaining trust" is actually the whole point of the existence of standards. Standards and standards-based processes provide us with the necessary yardsticks to understand what we are seeing, analyze how that compares to our expectations, and act accordingly. Standards, then, are a useful tool for building in transparency both during the software-development phase and the adoption, use, and maintenance of that software.

1.13 The Relationship between Code and Documentation is at the Core of the Marriage, Even after a Shotgun-Wedding

Only bureaucrats care about documentation, right? Wrong. Your regulators rely on documentation and even your customers will need some practical documentation to be able to adopt, trust, and use your software. Documentation is not something extra someone else creates for your software, it should be the shadow that your software brings with it every time it enters a room as illustrated by Figure 1.13. That said, superfluous, unnecessarily verbose, and self-centered documentation is something to beware.

The most basic way of checking whether your trust in someone is warranted or not is to ask them to explain their actions and compare that with what you know of their actions. Columbo, Hercule Poirot, Miss Marple, and just about every Tom, Dick, and Harry P.I. out there rely on this. This is also the way certification of those actions takes place: it's primarily a meta-level exercise based on what you document, both in terms of documents and records, with a supporting live component inspecting what you are doing. Similarly, the relationship between your software code and software documentation is mortally entwined: they are intended to be two sides of the same coin. If this is the case, great. If not, you may have a problem.

When you talk with some programmers, they almost instantly give off airs of hating documentation. Documentation is something almost dirty, demeaning, and certainly unnecessary to the beautiful code they could

Figure 1.13 Code and documentation should have a symbiotic relationship.

be writing. The cliché is that the more virtuoso a programmer is the less they want to explain how the magic in their code works. It should be self-evident. No one wants to plan, comment, review, or especially document the work they have done; the work should speak for itself. Only the work does not speak. It does not answer the questions regulators, B2B customers, and potential users have on their minds. For this reason, it is safe to assume that at least 50% of the software-development work in a regulated context will consist of work on more than just the code itself. Without the code the documentation does not matter, but without the documentation the code does not matter either – and if the code does not match the documentation you are in a world of hurt.

We too are programmers at heart – or were once before the matrix of it all swallowed us up and did its best to turn us into grim-faced suits with earpieces. Programmers – like us – often have strong opinions about how the code should look, how it should be composed to run as efficiently as possible, and especially on what is unnecessary to it all. This occasionally leads to some head-scratching and heated conversations when later deciphering code to explain it or perhaps reuse it in the next project. The use of a good programming style guide may seem superfluous or even blasphemous to some programmers, but it is a surprisingly powerful tool for meeting regulatory requirements and avoiding silly mistakes (or common defects as IEC 62304 might call these). A style guide coupled with consistent peer reviews is a great start to developing quality software, and as a solution one of the easiest to sell most programmers on. In a peer review, the previously defined style guide now acts as review criteria for the assessment and works to improve consistency and quality – and it does so in a way that can be demonstrated to match with the company's criteria.

This book will help you triage your documentation needs. Based on the discussion between these covers you'll be well-equipped to make decisions on what documents, reviews, and checks are truly worthwhile and which ones are just lengthening the to-do list.

1.14 Evaluation, Review, Verification, and Validation

> *This is the holiest of the trinities introduced by either of the two standards, at least outside of the seeming Bermuda Triangle of the patient, the user, and the customer – all of which may be the same entity or multiple different ones. Both standards place expectations*

on the ways you check up on performed work, and the negotiation of these requirements together holds the promise of a lean process. Figure 1.14 illustrates the complementary views offered by the standards.

The two standards, IEC 62304 and ISO 13485, look at the same thing – software in this case – from two different angles and want to ensure it all makes sense according to their points of view. IEC 62304 logically looks at the forest and the trees that make up the forest. ISO 13485 equally logically looks at whether your forest meets the requirements you had for a forest and the expectations your customer has for a forest. This inspection is to take place via testing on many levels as IEC 62304 sees it, but also for the purposes of verification (what you wanted) and validation (what your customer needed) as ISO 13485 expects. Luckily, both standards are in agreement on what is meant by the term review.

As a result, you can combine the requirements from the two standards in a great number of different ways. This is not unlike a Rubik's Cube where the colored squares somehow represent the requirements of each standard. You may decide to look at one aspect, or side, at first, but you will ultimately need to figure out what arrangement of reviewing, testing, verification, and validation makes the most sense for the entirety of your operations.

The terms evaluation, review, verification, and validation have distinct characteristics, but they are also sometimes used interchangeably or with an undefined overlap. For the purposes of developing medical software, the

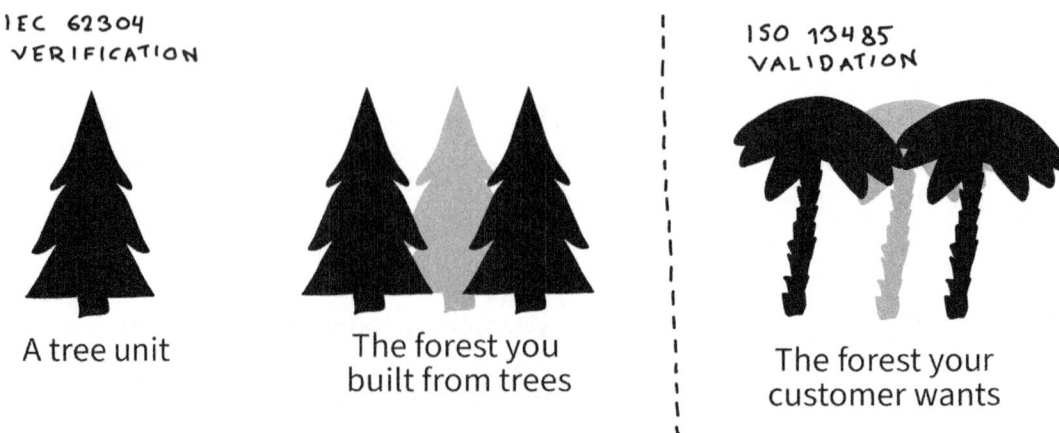

Figure 1.14 Verification and validation have similar goals, but different points of view.

definitions may come from either standard, the ancestral ISO 9000 behind ISO 13485, or in the absence of a definition in any of these standards, from the generally accepted dictionary definition. Based on this lineage the following key terms are defined:

- **Evaluation**

    This term is not defined by ISO 9000 or ISO 13485, but it is defined by IEC 62304 (Clause 3.7) as the systematic determination of the extent to which an entity meets its specified criteria. The ISO 9000, too, uses the term in several contexts (e.g., Clauses 2.3.3.4, 2.3.4.4, and 3.11.9) to indicate comparison against criteria. Thus, it is fair to conclude that an evaluation requires previously set criteria before it can take place.

- **Review**

    This term is not defined by ISO 9000, ISO 13485, or IEC 62304. Thus, the standard dictionary definition of assessing something with the intention of acting on any discovered shortcomings applies here. In other words, activities to assess verification or validation may also be said to be reviews. In contrast to an evaluation, no previously set criteria may be needed in a review.

- **Verification**

    This term is not defined by ISO 13485, but it is defined in detail by ISO 9000 which is thus the definition also required by ISO 13485. In ISO 9000 Clause 3.8.12, the definition of verification is given as confirmation through the provision of objective evidence (i.e., evidence on the existence or verity of something) that specified requirements have been fulfilled. This evidence is noted to come from, for example, inspection, measurement, testing, alternative calculations, or a review of documents. Note also that the term "specified requirement" is tied to a requirement stated in the "documented information" by ISO 9000 (see Juuso 2022) although the term "requirement" would have a much wider definition. Further notes in ISO 9000 explain that verification is sometimes referred to as a qualification process and that the word "verified" is used to designate the corresponding status. IEC 62304 Clause 3.33 agrees with the definition made by ISO 9000, reiterates the above note on the use of the term, and chooses to explain that in D&D the process of verification concerns the examination of the result of an activity for conformity with the requirements for that activity. The reference to D&D is an oddity in the sense that this is the only time the body of IEC 62304 uses the term, while in ISO 13485 it is very common.

- **Validation**

    This term is not defined by ISO 13485 or IEC 62304, but it is defined in detail by ISO 9000 which is thus the definition also required by ISO 13485. The definition of validation is given as confirmation through the provision of objective evidence that the requirements for a specific intended use or application have been fulfilled. Note here that the term "specified requirement" is not used, but instead, the generic term requirement is preferred allowing for a much wider base of analysis. A note to the definition offers, for example, test results, performing alternative calculations, and reviewing documents as sources of that evidence. As with verification above, the word "validated" is said to be used to designate the corresponding status. Interestingly, the use conditions for validation are here allowed to be real or simulated.

Keep the above definitions in mind if you find yourself wondering what a review is, what additional requirements are imposed for verification and validation, and how the three are related. Remember that all reviews should meet the requirements stated in ISO 13485 (see Section 5.9.1), but additional requirements from both standards apply to your body of verification and validation work (see Sections 5.9.2 and 5.9.3).

The most foolproof way of meeting all the requirements is to sandwich the activities by first seeing to IEC 62304 reviewing, testing, and verification requirements in a way that is forward-compatible with ISO 13485 expectations, and then consulting ISO 13485 to perform a summative assessment and ensure nothing is left behind. In this book, we adopt this model and run these activities first according to IEC 62304 and then look at the output review, verification, and validation before moving ahead with the release.

Alternatively, you could take the verification, and, at least to some extent, the validation requirements ISO 13485 has and infuse them into the testing activities of IEC 62304. Clause 7.3.2c of ISO 13485 seems to suggest this by requiring that your D&D planning documents the verification, validation, and design transfer activities appropriate at each D&D stage. This might lead to the most high-performance process where the output was the closest to being released with the minimum amount of after-the-fact checks needed on top of the IEC 62304 software-development process. Most likely, though, you would still end up with partially fused verification activities and a somewhat separate validation activity – especially as validation is required to be performed using a representative product (e.g., initial production units) and not some incremental development candidate still evolving on several fronts. Both verification and validation activities are also required to observe any needed connections to other devices.

These activities are discussed in more detail throughout the book, but especially in Section 5.9.

1.15 How Many Reviews Is Enough?

Are you an effective team? That is the question posed to Tom Cruise and Victoria Riseborough every morning by their boss in the film Oblivion. *This might be the leanest example of checking up on previous work and getting a feeling of the comfort and confidence level of your employees. The medical-device standards require you to ask a few more questions on the outcome of the work done during each stage, whether that is on the level of planning your development project or reporting on its outcomes. The number of reviews is often a case for some deep thought as Figure 1.15 illustrates.*

A central feature in all standards, and all planned activity really, is that the plan for doing something must be approved before that something is done. Coming up with a plan once you've already done something is self-evidently silly and you will be fooling yourself if you think you have done more than recorded the path you happened to take. This does frequently happen out there, though. The Plan-Do-Check-Act model invented by W. E. Deming in the 1950s – although he emphasized studying instead of checking – is invoked in the ISO 13485 standard as generally accepted good practice: you plan, you do, you check whether the actions and outcomes meet your plan, and then act accordingly. This should be your default mode of addressing just about anything in your business as time allows. But how does this translate to medical-software development?

ISO 13485 and IEC 62304 set some expectations for the activities you need to carry out. It may occasionally be debatable whether these activities need to be carried out according to a predefined process or a plan, or just performed according to your general arrangements, but it is never acceptable to not take responsibility for the results or somehow check on the output produced. This applies equally to the requirements you set at the start of a development project and the outsourced components perhaps going into your product through your assembly line. Here we are talking of a key type of evidence you will need to run your business and keep churning out the

Figure 1.15 "Review vertigo" may lead to more fatigue than attention to detail.

products your revenue depends on, but we are also talking about a key type of evidence you will need to show to your auditors whether those are certification bodies, notified bodies, or your B2B customers auditing you as a supplier (see Juuso 2022).

The question of how many reviews is necessary is very much up to you. Performing no reviews is not acceptable in any context other than a fairytale. Performing one review might be theoretically possible as long as you lock your plan documents in some airtight way and are happy taking a risk that your later actions then either meet or fall short of these plans and any acceptance criteria defined therein. In practice, this would be foolhardy and might fall apart as a solution the moment you would need to adjust your plans. On the other hand, performing a gazillion reviews is not feasible or probably beneficial either. So, the question remains, how many reviews might be optimal?

In this book, we look at the stages of software development in-depth starting from Section 5. The stages therein will provide you with a solid framework for structuring your activities and identifying the plans that need to be approved before moving on to the next stage. Within the stage too there are activities that you will want to plan before executing them. In some cases, it will be enough to refer to your standard operating procedures (SOPs) and acknowledge that these will be followed in your activities. In other cases, you will want to address project-specific aspects in planning some activity, and thus modify your general procedure within the limits your SOPs allow.

In general, the minimum level of specific review steps to address in your development model is three. These are:

- **Design Input Review (DIR)**
  This is where you assess the status of your inputs, especially the software-system requirements you have formulated based on all the information presented to you and available to you. Here you will have to listen to your customer, the applicable regulatory environment, and anything else you should know before getting started with the development project. The direct harm from fumbling with requirements will be that you waste resources in your project, but the ultimate harm may be that you misunderstand some customer requirements and create a less-than-perfect or potentially dangerous product. The DIR is discussed in more detail in Section 5.2.
- **Design Output Review (DOR)**
  This is where you assess the output of your design and development work. The objective here is primarily to check that your outputs and deliverables (incl. documentation) are as you planned. This review sits conveniently at the place where IEC 62304 is mostly finished with your product-development work, although software maintenance, risk management, and problem resolution will, for example, always continue to be topical. The DOR is discussed in more detail in Section 5.8
- **Design Transfer Review (DTR)**
  The last of the three big reviews is mostly a topic for ISO 13485. Here you must ensure that you are happy moving your software from development to production. The DTR is discussed in more detail in Section 5.8. This review may sometimes be called the Design Readiness Review (DRR) or even the Transfer Readiness Review (TRR), but the intention is the same.

The above-listed reviews are what you are expected to have a good handle on. It might not be uncommon to have half a dozen or even close to a dozen different reviews marked in your development model. The standards do not require any specific number of reviews or even specific names for the reviews, but they do want to see you plan and assess your work.

The specifics of what deliverables are reviewed in what review, or especially what lower-level reviews these major-ticket reviews may be based on, will vary. By going through the individual development stages later in this book you will be able to fashion your optimal process and frame that discussion in the terms expected by outside stakeholders.

A poignant question related to the big reviews is, "When should they be redone"? The simple answer is when the reviews may need to be updated for one reason or another. If the DIR defines input requirements but identifies some set of requirements as cumulating over the development project, or the risk-management file as iterated over time, does it then need to be redone if a new requirement is added or the matrix of identified risks is iterated as more is learned over time? What if a software change leads to a clarification of the wording of some label implicated in risk control? The answer is that the reviews should be revised any time they no longer accurately reflect the software-development work. If on the other hand, the change-management process arrives at a conclusion that the change does not affect the reviews, these probably should not be revised. The change-management process taking effect after the release of the product, if not before, should already cover the necessary reviews and thus be enough without redoing DIR, DOR, and DTR in a simple case.

1.16 How Much Testing Is Enough?

This is a question that almost invariably comes up at one point or another during the examination of any medical-device development path. Less testing naturally equals savings in the short term. If you're not careful, however, this time and money may be borrowed from the future, and the interest to pay may be steep when it needs to be paid. The purpose of testing is not to tick a box, but to have built up all the necessary confidence to either tick or not tick that box. This is not unlike walking out in front of a live audience after testing your standup skills on your friends, as Figure 1.16 attempts to point out.

Figure 1.16 Testing is required to trust that it all works.

The answer to the question of what is enough is infuriatingly simple: it depends on your product, its requirements, and how comprehensively your tests look at meeting those requirements. If there are any significant omissions or gaps in any of this, it will be difficult to make up for any shortcomings in one area with the extraneous testing you may have done in some other areas – so it is even possible to do both too much and too little at the same time. Before you throw this book at the wall, let that sink in. It means that you get to decide what needs to be tested for you to feel confident that all is as planned and expected. Now, before you get too arrogant you should remember that the requirements also include all the relevant expectations from regulations and standards, in addition to the features you and your customer jotted down on that napkin. Remember, too, that your regulatory authorities need to then also agree with your reasoning here.

It is thus possible to approach testing from two opposite angles that despite their differing points of view both have the goal of producing a better, safer product that meets the expectations. Firstly, IEC 62304 looks in-depth at the requirements and the software items built to those requirements, starting from units all the way up to the integrated software system. Secondly, ISO 13485, and regulations in general, look at your whole set of requirements (incl. applicable regulations and customer expectations) and your evidence for meeting those requirements with your final device. The requirements for testing from these two standards can be summarized as follows.

Starting from the point of view of the finished product, ISO 13485 expects the following:

- You have, in due time, defined what is required for the testing of the product and its acceptance criteria
- You have completed that testing prior to releasing the product and you now maintain traceable records to provide the related evidence
- The records include the identification of the testing equipment and, at least in the case of an implantable device, the identities of the testers. For the purposes of traceability and reproducibility it will make sense to retain a bit more information than required here, but the choice will be yours on what information to retain for these purposes

From the point of view of building that product, IEC 62304 expects the following:

- The software requirements are expressed in a form amenable to the creation of test criteria and the performance of tests (see Section 5.2)
- Testing is done to verify the implementation of software units unless this verification is performed by other means (see Section 5.5)
- Software-integration testing (see Section 5.6) and software-system testing (see Section 5.7) are done either separately or together
- After testing, all requirements have been tested or otherwise verified

Note that IEC 62304 expects even legacy software to have, as the bare minimum of deliverables identified by its gap analysis, a set of software-system testing records (see Section 4.4.2). See the referred sections for notes on how the software safety classification may influence the above-listed expectations.

These two standards are hard to avoid if you are reading this book and working on medical software. Beyond the two you must take a long hard look at your product to know what you need to test. Knowing if you are developing code for a pacemaker, an electronic health-record system, or a fitness tracker will affect your consideration here. So too will the user population, the use environment, the intended use, and everything else you think you know about your product and its future use. These characteristics will also affect your interpretations of the two standards already mentioned, of course.

Usability (see Section 12.5) is often a big-ticket item here and it may feed into your clinical validation activities sooner or later. The main standard for usability is IEC 62366. Another usual suspect is electrical safety based on the IEC 60601-1 family of standards. A recent important extension of safety is cybersecurity, which does not yet have a convenient single standard to refer to (although see Section 12.6) but is on everyone's lips from the regulatory authorities to the users. External testing services are available for each of these and much more.

For basic testing work all that is required is that you roll up your sleeves and get to work. You will still want to ensure people are not approving their own work, but mostly this is an in-house issue. For some other types of testing, such as cybersecurity testing, the involvement of an outside body is strongly implied. It is also worth pointing out that anytime your regulatory authorities see you not clear of the height they expect, whatever the topic of the shortcoming, you will be required to do more work to try the jump again or stay on the field. At some point they may lose faith in your abilities and either ask you to start again or ask you to bring in an external expert to assess the status. An example of this would be external testing against the IEC 62304 standard which may not be needed if your processes and documentation obviously meet the requirements but will be required if they don't.

Whether or not you are compelled to use external testing providers, you will still benefit greatly from having an outside expert come in to assess your processes in a lower-stress context than an actual device inspection and provide you helpful advice on how to both improve and streamline your processes. If you chose this expert wisely, by sticking to accredited testing providers, this work will be time away directly from your ultimate regulatory audits. If your choice is hard to justify, your notified body may throw out the testing work, possibly audit the testing body, redo testing in-house, or require you to have it redone. All of this will cost time and money.

Thus, testing is primarily an exercise in building up the required confidence in your own product. Once you are confident, you are done. The notified body will then test your confidence by asking about the testing you have done and see if it all holds water. This too will be a test, but it will not be a part of your testing activities per se.

As a final anecdote on why testing is such an important and often troublesome area, consider the fictional requirement of, "The device must read input from a keyboard". Here if the user types in "ABCD" and the program accidentally reverses that while reading it into "DCBA" the requirement is still met. No user would be happy with that outcome, but technically the test and its ill-conceived test criteria could be satisfied. Think about this when you work on defining tests and don't just go for the quickest answer. Defining tests is not just a chore to do – and fixing badly worded requirements discovered during the definition of tests will cause headaches of its own.

Testing is often a source of problems when performing IEC 62304 assessments. Defining requirements that are stated within this realm and that don't verge on the abstract or the superfluous, will help keep the later testing work manageable. Occasionally it happens that requirements are later discovered to be untestable in practice – and occasionally it even happens that this minor detail has purportedly not prevented the conducting of tests which are then well-documented despite not being possible. It goes without saying that both are occasions where conformity with the standard is called into serious question. Make sure you approach tolerances and acceptance criteria with all care – and do so before you conduct the testing itself.

Testing is usually a planned activity where the testing, test cases, and acceptance criteria are set during the planning stage. The key question is often not how many tests are performed, but by whom the tests are performed. Letting programmers test their own code is a natural part of development work, but in terms of conducting final testing they may be too lenient towards their own work or too familiar with the correct use of their software to tease out any issues. In recognition of this, the more hardware-oriented standard IEC 60601-1 Clause 14.10 even goes as far as to request documentation of the appropriate level of independence for the personnel performing verification activities (which may include testing). IEC 62304 does not make a similar requirement, but nonetheless using testers independent from the developers is known to produce good results.

1.17 Fail and Pass

An IEC 62304 conformity assessment is probably on your roadmap to releasing medical-grade software. You'll want to pass the assessment rather than fail it, which are the two choices shown in Figure 1.17. No one sets out to fail their conformity assessment, of

Figure 1.17 To pass or to fail, that is the Shakespearean question in software.

course, but understanding that there are some requirements where shortcomings will cause an immediate failure, and others where a failure may just need to be addressed in due time, may help you pass the assessment in a more straightforward fashion. This lesson points out a few of the "fail fast" requirements built into the standard.

In the IEC 62304 conformity assessment, your software-development activities are evaluated against each requirement of the standard, one at a time. The status of each requirement is then evaluated as a verdict using three possible values.

- **Not applicable (N/A)**
  Meaning that the requirement is not applicable to your operations. A requirement may not be applicable to your case based on the software safety classification (e.g., the requirement only applies to classes B and C and your software is class A) or because it has been labeled "as appropriate" and you have a documented rationale for why it doesn't apply (e.g., your software does not provide any user interface and can't therefore meet Clause 5.2.2f requirements; see Section 5.2). Note that the rationale for the applicability of the requirements will be recorded in the results/remarks section of the official test report.

- **Pass (P)**

    Meaning that your operations meet the requirement. For example, a verification plan is expected for the software, and the manufacturer has provided a plan that also meets all the content requirements for such a plan. The requirement will thus be judged as passed.

- **Fail (F)**

    Meaning that your operations do not meet the requirement. For example, a verification plan is expected for the software, but no such plan is found.

In practice, the most challenging part of the assessment is figuring out if the manufacturer meets or does not meet a requirement that applies to them. This watershed point is occasionally the source of some passionate debate. This is also an area where the standard falls a little short as it does not provide any guidelines or acceptance criteria for when a requirement may be judged as passed and when it must be judged as failed. Instead, the standard describes the ideal situation and then relies on the expertise of the assessor to determine what is close enough in the real world. As a result, the acceptance criteria may occasionally be subjective and dependent on the assessor.

There are, however, some requirements where a failure will cause an immediate stop to the conformity assessment. These fail-fast requirements are mostly addressed in Clause 4 of the standard (see Section 4) and address fundamental expectations of the standard. The top examples here are:

- Software development work must take place within an ISO 13485 quality management system
- Software development work must incorporate ISO 14971 risk management

A few typical lower-level fails that are also easy to spot might be:

- One of the key planning documents required by the Software Development Plan does not exist (see Section 5.1)
- No acceptance criteria have been defined for software testing (see Section 5.1)
- The software risk-management activities do not cover the use of SOUP components although these are used by the software (see Section 7.1)

The above examples are somewhat Boolean in nature – or Shakespearean as in to be or not to be. The previous shortcomings may be simple to spot, but that is not the case with most other requirements in a real-world context. Examples of shortcomings that will be harder to spot and to assess might include:

- The verification of requirements does not adequately address the disambiguation of requirements, their mutually contradictory nature, or their readiness to be tested (see Section 5.2)
- The safety and performance of the medical device have not been adequately tested (see Section 5.7)
- The risk management of the medical device does not cover all relevant risks (see Section 7)

The standard also covers some challenging requirements that often benefit from a synchronization of expectations between the manufacturer and the assessor prior to the assessment. Examples here include the handling of SOUP components based on the software safety class, erroneous specification of functionality as a potential cause of harm (Clause 7.1.2a; see Section 7.1), and what it means to reverse adverse trends during software problem resolution (Clause 9.7b; see Section 9.7).

As mentioned earlier, the standard does not contain nor is it accompanied by guidance on what solutions may be considered as meeting the requirements of the standard. The non-binding Annex B of the standard does attempt to provide some further friendly guidance on the requirements and what they actually mean. The description in the appendix is brief, non-binding, and void of sufficient examples to truly foster a uniform understanding of the requirements across all stakeholders. Hopefully we will at some not-too-distant stage see a good technical report or other guidance document introduced to provide some practical framework for acceptance criteria and a set of actionable examples to assist in the assessment of each set of requirements.

1.18 One Step Forward, Two Steps Back

The standards are very good about requiring that you always keep everything up to date, but how do you accomplish this without constantly going back to update documents for every small thing? One

Figure 1.18 Maintaining forward momentum may at times be difficult.

step forward and two steps back is not a good model for a process, as Figure 1.18 illustrates.

In our earlier book focusing on ISO 13485 quality management (Juuso 2022) we noted that updating documents to obtain new timestamps is just all kinds of silly. Similarly, updating plans to keep track of progress and the produced deliverables is counterintuitive – except for when your plans really do change. The software-development life cycle sketched out by IEC 62304, too, is littered with such back-and-forth connections between deliverables and development stages. As a result, a development process written merely to comply with the standards may be far from lean if you are not careful. If you take an analytical approach to what the underlying goals of each activity are and how best to ensure those goals are met, you will find that the standards do lend themselves to the design of lean processes.

Arriving at lean processes that move ahead reliably and predictably is no small feat. The historical and organizational baggage of your organization will weigh in heavily here, but so too will any best practices and practical conventions you have developed over the years. If you are starting anew or looking to overhaul your processes to something leaner the following design goals may help you get there.

- Take a long hard look at the processes you are accustomed to, and how these compare to what is expected by the standards and regulations. Don't just create something totally new that is so revolutionary no one will want to learn it or follow it. A QMS is not a folder of aspirations that you bring out for the auditor to marvel at, it should be your optimal way of life: your way of doing things so that you meet all the requirements.
- Define your generic, overall processes in the SOPs and refer to them in any plans you need. This should include your basic approach to testing,

verification, and validation, too. Having one central location to refer to will simplify the maintenance of links with any related QMS processes (incl. staff training and customer communication), but it will also simplify the actual running of those processes.
- Add project-specific details in a software development plan. This should include how this project is expected to be different from the generic project discussed by your SOP. Your project must still fit within the parameters allowed by your SOP, of course.
- Don't jump ahead to micromanage things. Instead, record important things to address during a later stage and iron those details out at the appropriate time. An example here might be to attempt defining executable tests for all the requirements when you are not yet sure of your final set of requirements – instead verify your requirements as expected by the standards (see Section 5.2) and then later define the tests before testing takes place.
- Avoid copy-paste. If you find yourself entering the same information in several documents, consider using references instead and take a long pause to figure out when the particular piece of information or instruction should actually be chiseled in stone and made available.
- Don't update documents just for the sake of new timestamps. Just don't do it. It makes no sense and causes more issues downstream as people will have to figure out what has changed and how it affects their work.

This book will help you fashion a streamlined process for your software-development activities. The discussion from Section 5 onward will increasingly discuss the requirements of the two key standards, ISO 13485 and IEC 62304, and guide you toward making the right choices for your organization. The ultimate expectation of any industrial reading of the standards is that by following the map they provide you will be sailing ahead towards your destination in a smart, efficient, and predictable way making at least discernible constant progress. Not having tailwinds to carry you all the way may be expected, but rowing in the wrong direction, or backpaddling just to record your recent progress, does not make much sense.

1.19 Continuous Improvement

On the previous note of maintaining forward momentum, another perhaps related essential concept in the standards is the notion of

44 ◼ *Medical-Grade Software Development*

assessing where you are and then moving ahead in the right direction to improve the status quo. A QMS should be getting better over time, as Figure 1.19 illustrates.

Years ago, in the heyday of the American automotive industry, a high-up executive was quoted as saying that to have more than one feature which sets you apart from your competitors is wasteful. The implication here was that to excel in more than one aspect would be superfluous as the customer would already buy your product based on the first aspect. This notion appears drenched in hubris and is, today, antithetical to what modern quality management and standards like ISO 13485 want to see.

It does not make sense to constantly adjust processes, but to not measure your processes somehow and to not know both where you are and where you want to be is foolish. The father of modern quality management, W. E. Deming, pointed as much out when coming up with his 14 points to guide company management through quality management. You should, however, always be on the lookout for areas of possible quality improvement, and the ways you could then implement to achieve those improvements. This vigilance and improvement should be continuous and over time lead to vastly superior processes.

The ISO 13485 standard in particular wants to see your whole organization working to improve upon the status quo. Small steps applied over a

Figure 1.19 Your path into QMS should have a rising crescendo.

long time will get you to a much better state of affairs over time than a few dramatic and perhaps ill-conceived actions. This applies to the management of your organization, the management of quality throughout your organization, and it also applies to how you can approach the development and fine-tuning of your software-development processes. What you have now does not have to be perfect, but it does have to be good enough and worth improving.

1.20 What Is SOUP?

> *Soup du jour? SOUP is a special concept in software development and one which is a little more than just the chowder served at your local establishment today. SOUP and soup have more in common than just the letters, as Figure 1.20 illustrates.*

Admit it, you instantly thought of that Andy Warhol painting of Campbell soup? That's not a terrible analogy here, as it turns out. Much like the painted can of tomato soup in the picture, which you can only imagine contains tomato soup inside, you have to trust that the contents of some piece of outsourced and possibly expired software does what it was intended to. That's what the standard understands with the acronym "SOUP" which stands for Software of Unknown Provenance.

SOUP is software that is generally available, developed for some other purpose than going specifically into your product, or the development

Figure 1.20 SOUP or not, you'll need to know what's in it.

process of which you might not have all the adequate records. The whole software system itself can't be SOUP, but it can be what is known as "legacy software" with somewhat similar repercussions (see Section 1.22).

SOUP is a fundamental concept to software development and one which you may already be familiar with. The topic will be brought up frequently in the discussion in the following sections.

1.21 Configuration Items and Controlled Changes

> *It is intuitively only appropriate that something painstakingly designed, reviewed, and approved can't be changed in any haphazard way thereafter. If changes were not addressed with rigor, the best of devices could be undone over time and at the very least their pedigree could be called into question. Part of being able to control change is to have mechanism for measuring and labelling it, as illustrated in Figure 1.21.*

Having periods of blackout, or uncharted history, is unacceptable from an auditing point of view – unless we are talking about legacy software (see next lesson) – but the blackouts would also make it more difficult to progress with the design of your products later on with any degree of certainty.

Figure 1.21   Configuration covers both versions of your product and what is in it.

The first step to controlling changes is to identify the states across changes, i.e., the versions of the item changed. This is done through what is called configuration items by IEC 62304. The process for identifying such items and controlling changes to them is discussed in detail in Section 8. The choice of what items to control is on one hand simple: you want to control all the parts of your medical-device software. On the other hand, it may not be quite that simple as the documentation of that software and the data used by the software may also need to be controlled. In general, any changes that may affect the conformity of your device with its requirements need to be controlled regardless of if these changes apply to the source code, documentation, models, data, or databases related to your software. See Section 8.1 for more discussion on the use of configuration items.

While on the topic of configuration management, it is pertinent to discuss the related topic of traceability, which is a core concern in medical devices. Regulations and the ISO 13485 standard, too, place requirements on how you enable the identification of medical devices, and how you trace discovered or otherwise investigated issues to their potential sources. Thus, traceability is something you are required to do, but done right it can actually be a competitive advantage for your organization: something which allows you to investigate issues and answer design questions in a smarter, quicker way during development, too. In this vein, and the part of most interest to IEC 62304 in this context, is traceability between various D&D deliverables or items, such as:

- System requirements, software requirements, software-system testing, and risk-control measures implemented in software
- Software items and hazardous situations [B C]
- Software items and risk software causes [B C]
- Risk software causes and risk controls implemented in software [B C]
- Risk controls implemented in software and their verification [B C]

In the above, superscript lettering denotes the software safety classes (i.e. A, B, and C) the traceability requirement applies to. Also note that, as per IEC 62304, the term software item is to be understood as software system, software item, and software unit as is appropriate in each context.

The D&D documentation (incl. systems used to provide traceability) must enable the above traceability as required by ISO 13485 and IEC 62304. Additionally, it should be noted that risk management as per ISO 14971 has further requirements on traceability between risk-management activities and deliverables (see Section 4.1.4.5).

1.22 Legacy Software Has a Dedicated Pathway Through IEC 62304

The term "legacy" perhaps conjures up images of dynasties and foundations keeping alive the heritage of the tycoons who established them decades ago, but in the standards, it refers to something which is not easily discarded even if it isn't the focus of intense interest today. Doing away with such software might shutter some devices and services users still rely on but updating everything to modern standards would require an oversize investment on the part of the manufacturer. To keep such software humming on in a reliable way the standard provides a special pathway, as illustrated by Figure 1.22.

If that can of tomato soup was painted by you yourself or is by now only stocked by your private cellars, it may more accurately be called legacy software. The concepts of SOUP and legacy software are distinctly different

Figure 1.22 The IEC 62304 has a charted path to conformance for old software.

but nonetheless related. Neither necessarily meets the full IEC 62304 requirements.

Legacy software is a term used for software built earlier that we still want to use but that has not been developed according to IEC 62304. The 2015 amendment of the standard addressed just such legacy software and provided a new abbreviated pathway to achieving conformity with the standard. Just like the software safety classification mentioned above (see Section 4.3 for an in-depth discussion), classification as legacy software means that how the various clauses of the standard apply to the software is changed. This matter is addressed in detail in Section 4.4, but for now, it's enough to know that legacy software has a special pathway of its own through the standard.

The point of legacy software is not that you are too lazy to upgrade your code, or that you want to squeeze out the last drops of value from an aging piece of software. That would be a dubious proposition, perhaps similar in outcome to what happened when a good concept from the 1960s – a low-ground-clearance airplane that could be loaded and unloaded from the ground without any special equipment – was set to go out on the pastures but was instead given one last tweak – in this case to accommodate larger modern fuel-efficient engines: the whole veered off its goal and crashed to the ground. The point of legacy software is to keep something available that you know works, not tweak it unnecessarily, and not rob the users of the benefit they have been enjoying. Legacy software is not a great plan for the future, but it may be a good way of keeping around something already widely used.

As a curiosity, also note that what the EU MDR defines as a legacy device is again different to what IEC 62304 understands by the term. In EU MDR parlance, a legacy device is a device put on the market before the MDR legislation took effect, and that remains on the market after the date of application of MDR (EU 2017/745). If this sounds foreign to you, forget about it, as Anthony Soprano would say. If you are only now starting work on developing and releasing medical-device software this does not affect you.

1.23 Software Always Fails, and Other Software Risk Beliefs

Risk management is the topic of hundreds if not thousands of books written on management in general, and a good number on medical devices in particular. For our discussion in this book understanding the basics of risk management is enough. If you happen to be familiar with the ISO 14971 standard on risk management for medical devices all the better. Figure 1.23 provides a filmic reference for risky business.

Risk management is a complex topic best left to the ISO 14971 standard, and the countless books written on the subject, to explain. The IEC 62304 dabbles in risk management with good intentions, but not always successfully. Your ISO 13485 quality-management system most likely already contains processes for risk management so you should ensure these are observed and reused here as much as possible. The purpose of IEC 62304 is to make sure all relevant software-specific aspects are considered in your risk management operations.

The two most common beliefs when it comes to how ISO 14971 risk management is different for medical-device software than it is for other types of devices are that a) software always fails, and b) you can only use external, non-software risk controls to mitigate software risks. Neither is true, but as ideas to shake you awake to the importance of risk management with software both are valuable. These topics are further discussed in Section 4.3 when assigning software safety classifications and in Section 7 in the context of the full scope of your risk management activities.

Figure 1.23 You don't have to be Tom Cruise to find your "Porsche" in a creek if you ignore the signs.

For now, it is worth knowing about the most essential terms related to risk management. These are: hazardous situation, hazard, harm, and risk. The term "hazardous situation" (Clause 3.35) refers to circumstances wherein people, property, or the environment are exposed to hazards. A "hazard", then, is a potential source of harm (Clause 3.9), and "harm", in general, is some physical injury and/or damage to the health of people, property, or the environment (Clause 3.8). Finally, risk is the combination of the severity and probability of that harm (Clause 3.16). These terms will be discussed in much more detail in Section 7, and even before as appropriate.

1.24 Certification Bodies and the Notified Body

The concept of a notified body is a purely European concoction, but the various types of other testing and certification bodies you may engage to assess your operations according to a standard such as IEC 62304 is similar the world over. Figure 1.24 illustrates how certificates differ.

You may choose to ask a service provider to assess your software life-cycle activities and deliverables for conformity with a standard such as IEC 62304. No service provider will be able to issue you a certificate on IEC 62304 (see Section 3.2.1), but the results of the assessment will enable you to convince your stakeholders your pipeline is in great shape, and short of that, obtain expert advice on where to improve your operations. You may also be required by your notified body to have the assessment performed.

Your choice of service provider to employ here is wide and ranges from hired consultants and unaccredited certification bodies to accredited certification bodies also acknowledged by your notified body. As remarked on elsewhere in this book, the choice here is up to you, but it may mean the difference between having your notified body accept the assessment

Figure 1.24 Know that not all certificates carry equal value.

as is, wanting to audit your assessor themselves, or demanding that it is all redone. On the other hand, after a successful IEC 62304 conformity assessment, your notified body will be able to mark some questions off its list and concentrate on the remaining assessment preceding the clearance of your product. The certification body thus concentrates on, perhaps, a single standard like the IEC 62304 whereas the notified body must also observe other applicable standards and regulations and addresses, for example, clinical evaluation in much more detail.

In general, a certification body (CB) is an organization that is accredited to provide testing and certification services according to some specific standards in a particular domain of industry. The accreditation will be granted under some specific accreditation scheme, such as ILAC (https://ilac.org) or IECEE (https://iecee.org). Under the scheme, the national certification body (NCB) will license testing laboratories (CBTL) to conduct assessments under its authority and it may then issue certificates on passed conformity assessments.

No certificates on IEC 62304 can be issued under the CB scheme. Instead, under the IECEE CB scheme it is possible to certify a medical device based on the IEC 60601-1 standard and include a mention of IEC 62304 under the applicable collateral and particular standards via IEC 60601-1 Clause 14. On the certificate the IEC 62304 standard can, at most, be listed under the section for additional information. In absence of a certificate the final document from an IEC 62304 assessment is the test report, which you can then show to your stakeholders and the notified body.

Similarly, it may be possible to include IEC 62304 assessment as part of an overall ISO 13485 assessment or even to obtain a certificate of some sort on such conformity. In this case, the ISO 13485 certificate may include a statement that the design and development activities of the QMS adhere to IEC 62304 requirements for software development. The weight of such a certificate may be debatable and will depend greatly on the credibility of the issuing organization. A certificate you obtain from your uncle or from a traveling snake oil salesman may not be worth much, but an optional third-party certificate may carry at least some partial weight during your overall conformity assessment. This is a matter you may want to discuss with your certification body.

When in doubt, consider obtaining the services of an accredited testing and certification provider that is part of either the ILAC or the IECEE schemes. This way the verdict of the assessment will also be accepted by third parties issued in the CB scheme. Conformity assessment and regulatory approval are discussed in more detail in Sections 13 and 14, respectively.

1.25 Safety Is Relative

What do Butch Cassidy and the Sundance Kid, on the one hand, and the square-glassed pensioner in the animated film "Up!", on the other, have in common? Both go flying off to adventure, but it is probably safe to say that they have wildly differing expectations of what is to come. Figure 1.25 shows one way to mitigate dangers.

Rather mind-blowingly, safety is defined as freedom from unacceptable risk by IEC 62304 in its Clause 3.21, where the standard leans on the ISO 14971 standard for risk management. This makes sense, of course, but it means that a skydiver jumping out of a plane with an unchecked parachute and a granny sitting inside her locked house might both be considered safe

Figure 1.25 Marshmallows on cobra fangs might increase your sense of safety, but will it be safe enough?

provided that the expectation of acceptability is adjusted separately for each. The needs for insurance, then, should be the same for both people, but somehow the price, premiums, and deductibles involved probably are not equal in real life.

This realization lends new credence to the long-running debate on how the acceptability of risk is defined under applicable regulations – whether something mitigated to an acceptable level is also mitigated as low as possible, or vice versa. The debate is not just about semantics as it now represents the difference between setting acceptability at an appropriate level versus a devil-may-care level. If, for example, the criminal being asked if he felt lucky by Dirty Harry standing above him with a Magnum 44 had set the acceptable risk of miscalculating the spent ammo at $\pm$ 6 bullets he might have felt completely safe to gamble with his life. If he cared about his life this would have been a high-risk proposition. The reason that the definition of acceptability matters so much is not just that a higher risk gets you closer to danger, but that an unacceptable definition of acceptability may have you in significant constant danger. Let this thought linger as we move over to look at the two essential standards, ISO 13485 and IEC 62304, and their combined requirements next.

A minor note here is that IEC 62304 actually talks of both safety and SAFETY. The former is solely used in the context of the software safety classification while the latter refers to the above definition based on freedom from unacceptable harm. Technically safety, spelled with lowercase letters, refers to the common dictionary definition of safety but this is only used in the context of the software safety classification, and as the assignment formula for that classification is concerned with freedom from harm and injury (see Section 4.3), the difference between safety and SAFETY is academic.

Chapter 2

ISO 13485 AS THE BACKBONE OF IT ALL

Quality management will form the backbone of all your operations, including those centered around your software development and product-realization activities. This section will introduce you to the international ISO 13485 standard that forms the basis of just about all work on medical devices. Here is where you'll start putting together the T-800 endoskeleton of your model for medical-grade software development. In the next section we will then glue on the exoskeleton parts from IEC 62304 software development.

By now ISO 13485 is an old, trusted friend to much of the medical-device world. We have followed it to the bottom of the rabbit hole so deep and gone without competing forms of oxygen for so long that it is hard to imagine adopting another guide to navigate the global medical-device business. Maybe this could be called succumbing to the Stockholm syndrome, falling in love with your captor, or maybe it just means many of us see the standard as the logical expansion of the Helsinki declaration to the running of a medical-device business where human subjects are involved both within and without the factory walls. Either way, the standard is very much a part of the "way of life" for this area of business.

The standard is not perfect – see the earlier QMS book written by Dr Ilkka Juuso on the standard (Juuso 2022) for more discussion – but it is a compelling approximation of perfection. The ISO 13485 standard forms the backbone of so many critical initiatives in the field that you may consider it bedrock here. The Medical Device Single Audit Program

(MDSAP) alone has anointed the standard as the messiah of the global medical-device business, but also the US FDA and the European Union are increasingly hitching their wagons to the standard – cautiously, but purposefully. At a recent event, Dr Jeff Shuren, the director of the FDA's Center for Devices and Radiological Health (CDRH) put on a very positive spin suggesting American companies consider moving ahead with the transition to the standard now. European companies are already there with the advent of the EU MDR. This is all great news for a global medical-device ecosystem, where harmonization and the compatibility of regulatory solutions enable faster, safer, and more affordable manufacturing of medical devices. The companies will benefit from this reduced complexity and overhead, but so too will the patients, doctors, and other healthcare customers in the form of new innovations reaching users quicker and cheaper across the globe.

There may be future disruptions to this standard too as regulatory expectations evolve as a result of new more advanced and autonomous software products appearing on the scene, but nothing in the past decade has seriously challenged the contents, shape, or role of the standard. Ripping the standard apart and putting it back together now would be an act of international terrorism on the medical-device industry and the physicians and patients relying on medical devices and medical software around the world. Instead, expect accompanying guidance on the application of the standard to new relevant subfields of medical devices.

2.1 The Whole of the Quality Management System

The ISO 13485 standard covers the entirety of your medical-device operations via a comprehensive quality management system. The whole system is aimed at giving structure to your operations in a way that enables you to develop, release, and support medical-device products reliably.

The topics covered by the standard are:

- QMS documentation (incl. both documents and records)
- Corrective and preventive actions (i.e., CAPA)
- Monitoring and improvement of the QMS
- Infrastructure
- Human resources (incl. qualifications and training)
- Suppliers and distributors

- Auditing (incl. internal audits, external audits, and the management review)
- Communication, marketing, and sales
- Risk management
- Product realization

The topics touched on by the standard, but not covered in any detail, include the following:

- Clinical evidence
- Regulatory affairs (incl. product registration)
- Post-market surveillance

All these topics are discussed in our previous book on ISO 13485 quality management (Juuso 2022). In this present book, we are mostly interested in how software development can slot into the wider ISO 13485 quality-management context. The topics listed above are thus only addressed to the degree necessary here. For a more in-depth discussion of the topics please refer to our earlier book.

2.2 The Process for Product Realization

In the ISO 13485 standard, matters related to product realization (incl. development and maintenance issues) are chiefly a topic for Clause 7 on product realization. This massive clause comprises roughly 34% of the whole standard and could even be considered a standard within the standard. The clause is intended to preside over all your product-realization activities. It instructs subprocesses such as planning of product realization, customer-related processes, design and development (D&D), purchasing, and production and service provision. In this context, design and development includes planning, inputs, outputs, reviews, verification, validation, transfer, control of changes, and the file set. It is worth noting that the ISO 13485 standard does let the user of the standard exclude all or some subset of the requirements in Clause 7 if they have a valid rationale for doing so. It is thus possible that an organization certified to ISO 13485 quality management does not actually implement any of the requirements regarding the development or manufacturing of devices. An ISO 13485-certified quality management system may mean different things to different organizations, but the elements and processes they do implement should be compatible on a high level following the standard.

In addition to Clause 7, other clauses of the standard may relate to product-realization processes. This will often be the case as the purpose of the organization will likely be linked to the medical device it somehow manufactures or makes available to its customers. Typical examples of requirements covered elsewhere in the ISO 13485 standard but connected with activities you might undertake in the context of software development include tool validations, supplier management, and marketing communication.

Clause 7 is the most readily overlapping area of common interest between the ISO 13485 and IEC 62304 standards. The overlap will be analyzed in-depth in the sections to come, but here it is useful to already set up the general hierarchy of the court these standards keep. The patient is naturally the king in this regal setting, but it is the ISO 13485 standard that sets up many of the departments, professions, and pastimes of the castle. IEC 62304 makes use of this setting and goes into more detail on how the engineers, blacksmiths, and other tradesmen might then go about their business creating the deliverables asked of them. In other words, ISO 13485 provides the environment the work takes place in and many of the supporting services the work relies on (e.g., customer communication, purchasing, and the final provision of the product or service), and IEC 62304 adds the nuts and the bolts expected in between for software while relying on what the ISO 13485 has already said. Remember that having a QMS is not optional if you want to use the IEC 62304.

A rough sketch of the areas of common interest between the two standards, or the overlap of sorts between them, is represented in Table 2.1 below. The table is a simplified overview of the general lay of the land, and practice-related requirements, both big and small, are served out peppered all over each of the standards. An empty cell in the table does not therefore indicate that the standard would not have any wishes on the particular topic only that it doesn't dedicate real estate to discussing it. The discussion in the coming sections will address all of the topics in good time.

ISO 13485 sees product realization as consisting of design and development stages, the execution of which is planned in advance and supported by a range of other processes related to the customer, purchasing, and product provision. IEC 62304 shares the notion of planning and complements the structure of design and development stages but is less concerned with any supporting processes.

Table 2.1 A Rough Mapping between ISO 13485 Clause 7 and IEC 62304

Topic	ISO 13485	IEC 62304	See
Supporting processes	7.2 Customer-related processes	-	Juuso 2022
	7.4 Purchasing		
	7.5 Production and service provision		
	7.6 Control of monitoring and measuring equipment		
Planning	7.1 Planning of product realization	5.1 Development planning	5.1
	7.3.2 D&D planning		5.1
Inputs	7.3.3 D&D inputs	5.2 Requirements analysis	5.2
Design and implementation	-	5.3 Architectural design	5.3
		5.4 Detailed design	5.4
		5.5 Unit implementation and verification	5.5
		5.6 Integration and integration testing	5.6
		5.7 Software system testing	5.7
Outputs	7.3.4 D&D outputs		5.9.1
Review	7.3.5 D&D review		5.9.1
Verification	7.3.6 D&D verification		5.9.2
Validation	7.3.7 D&D validation		5.9.3
Transfer	7.3.8 D&D transfer	5.8 Release	5.8
Documentation	7.3.10 D&D files		11
Configuration management	7.5.8 Identification	8 Configuration management	8
	7.5.9 Traceability	8.1 Configuration identification	

(Continued)

Table 2.1 (Continued)   A Rough Mapping between ISO 13485 Clause 7 and IEC 62304

Topic	ISO 13485	IEC 62304	See
Maintenance	7.3.9 Control of D&D changes	6.1 Establish software maintenance plan	6.1
		6.2 Problem and modification analysis	6.2
		6.3 Modification implementation	6.3
		8.2 Change control	8.2
		9 Problem resolution	9

The design and development stages defined by ISO 13485 are planning, inputs, outputs, review, verification, validation, and control of changes. The standard also has a clause on the files used in these stages. In the list of stages, it is notable that a gaping hole exists between the input and the output stages. This is where IEC 62304 comes in to instruct what it means to be between inputs and outputs and what work is needed to proceed from inputs to outputs. The match here is complementary, if not quite plug-n-play.

In general, ISO 13485 provides the larger framework within which IEC 62304 can sit. Often the former provides the process definitions the latter can use to create the documentation and (other) deliverables required. IEC 62304 adds necessary information to processes and documentation (such as the software development plan) to capture the needs of software development.

Note that Table 2.1 essentially matches with the rough mapping given in Annexes C and D of IEC 62304 (see Sections 3.2.14 and 3.2.15). We have here opted to place system testing as part of D&D instead of as something only looking at its outputs (although that too is a defensible interpretation) and similarly taken the view that reviews, verification, and even to some extent validation activities can – and often are – fused into the development life cycle of software. The marriage of requirements from IEC 62304 and ISO 13485 is covered in-depth in Sections 5 to 9 of this book.

The ISO 13485 standard itself is addressed in great detail in the earlier book written by Dr. Juuso, *Developing an ISO 13485-Certified Quality Management System* (2022). The compatibility between ISO 13485 and IEC 62304 requirements is the topic of this book and will be addressed in detail in the coming sections.

Chapter 3

IEC 62304 AS THE FLESH AROUND THE BONES

If ISO 13485 quality management is the backbone of all medical-device businesses, IEC 62304 provides the exoskeleton around this to give you the necessary rigor for medical-grade software development. This section introduces you to the standard, and together with the following sections, helps you negotiate together all the pieces to shape your software-development activities in a way that fits snuggly with the expectations of quality management and regulatory requirements in the domain of medical devices.

As standards go, the IEC 62304 is a brief, but weighty piece of literature. The standard is not the prettiest one, nor does it play the nicest with the other kids in the standardization playground. In fact, the IEC 62304 has had a fairly troubled life being bogged down with revision woes for much of its existence. The standard is, however, widely used in the development of medical devices, and a de facto requirement as an extension to ISO 13485 quality management for any medical-device company engaged in the development of software. Your certification body will expect to see evidence of how you have considered the requirements posed by the IEC 62304 standard and how you have met these in your products and operations if any of these mention software. Depending on what level of billing you give software, whether your product is software or contains software, you may find these expectations need to be met via an in-house assessment or an external assessment by a third party the regulatory body can trust.

DOI: 10.4324/9781003454830-3

Bruised and battered, the standard continues to give us the yardstick against which all software development and maintenance operations are measured. That alone is reason enough to take it seriously.

3.1 The Past, Present, and Future

To understand the future, we shall begin by looking briefly at the past and the present. This will help us grasp the tectonic movements in the development of software that have reverberated through the industry in the past decades. Later in Section 12 we will point the juiced-up DeLorean even further into the future and reach into our crystal ball to discuss the phenomena shaping the near future of medical-software development in more detail.

3.1.1 Times of the Primordial Ooze

The rise of the integrated circuit, that ubiquitous brain-on-a-chip Lego-block of the electronic world, some time in the 1980s increased interest in and the development of software as part of electrical medical devices. At this time the circuits, also known as microchips, became commonplace in the design of physical electrical devices. The combination of the microchip and the software that it could run unquestionably brought about greater computational capabilities for the devices but also enabled some completely new functionalities for the devices. This small revolution led to greater emphasis being placed on the criticality of software to the safe operation of devices. Subsequently, work began on developing requirements for not just the devices themselves but also for the software embedded in the devices. At the same time, the manufactured devices themselves graduated from purely electronics-based devices, such as an RCA controller-based sound-level measurement device in the 1970s, to more modern Intel 8088-based architectures for patient monitoring devices around 1985.

The safety of electrical medical devices themselves has been regulated since 1977 by the first version of the international standard IEC 60601-1. This did not, however, address software in any way. The second edition published in 1988 and amended in 1995 cautiously mentioned software and the, at the time, upcoming collateral standard IEC 60601-1-4 specifically on software.

In 1996, the first true fruits of this work on software-specific safety requirements saw the light of day as the IEC 60601-1-4 standard was published as part of the IEC 60601-1 series of standards. Finally, in 2006 hot on the heels of the third edition of IEC 60601-1 came the first version of IEC 62304 which has since become a yardstick for medical software. All these standards were published by the International Electrotechnical Committee, that is still, among other standards, responsible for the IEC 62304 standard today. This series of standards created the foundations for assessing the safety of medical-device software. The original series of five standards addressed general requirements, the system, EMC, radiation, and software. This set has since been extended with a number of collateral and particular standards as shown in Figure 3.1. Two standards of the original set, system, and software, have since been withdrawn as is evident from the missing collateral standards 1-1 and 1-4 in the figure.

Part 1 of the series contains general requirements for basic safety and essential performance of medical electrical equipment. The other parts, called collateral or particular standards, supplement or modify the Part 1 requirements for particular types of devices. Part 1-1 and Part 1-4 on software have since been withdrawn, but Parts 1, 1-2 (EMC) and 1-3 (radiation) remain active in revised form. Also, additional parts have since been published on usability (Part 1-6), alarms (Part 1-8), environmentally conscious design (Part 1-9), physiologic closed-loop controllers (Part 1-10), home healthcare (Part 1-11), and emergency medical-service environments (Part 1-12). These standards may be of interest to you if your software product touches on any of these domains, but they are beyond our discussion here.

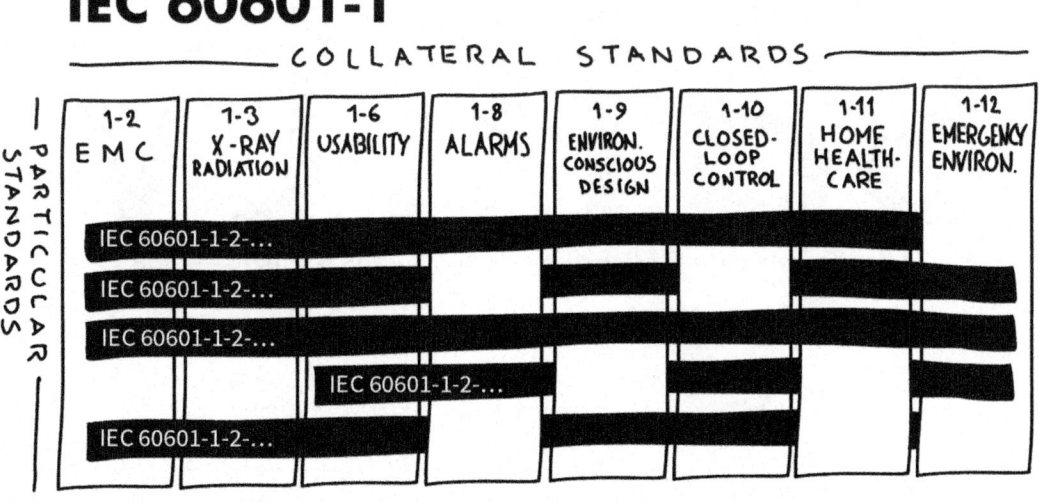

Figure 3.1   The collateral and particular standards of IEC 60601-1.

64 ■ *Medical-Grade Software Development*

As a package of standards, the IEC 60601-1 is still one of the most expensive items you may find yourself buying.

It is noteworthy that the scope of the standard was from the outset tightly coupled with the safety of electrical medical devices and the software embedded in those electrical medical devices. This was also the case in the collateral standard Part 1-4 on software, where the scope of the standard (its Clause 1.201) looked at the safety of programmable medical electrical systems (PEMS), which it saw as being either equipment or systems both containing programmable electronic subsystems (PESS) as illustrated in Figure 3.2.

Technological advances in the 1990s led to increased networking between devices and the utilization of server infrastructure to greatly boost the computational capabilities of devices. This development thus started to move the focus from a physical electrical device to the networks now used to store data, analyze it, and reuse it in later diagnosis.

Part 1-4 of the IEC 60601 series of standards, the collateral standard on software, was subsequently updated in the year 2000, but it was soon admitted that a new standard was needed to address the use of software in healthcare even in cases where no physical device was implicated. At the time, Europe was living with the Medical Device Directive (93/42/EEC), and the view started to emerge that software could be a medical device by itself as understood in the regulation.

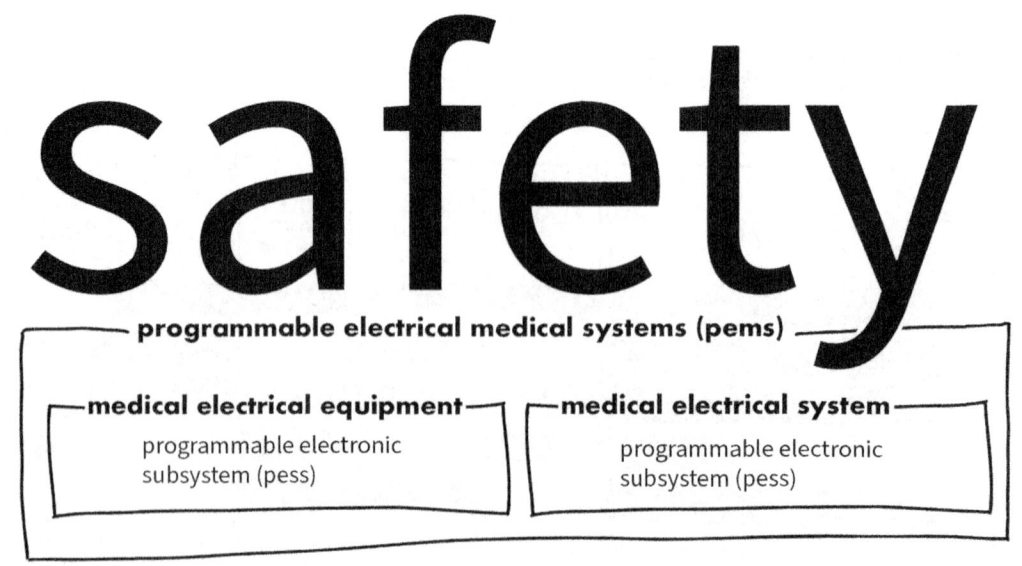

Figure 3.2 Scope of IEC 60601-1-4 on software.

Finally, in 2006 a new standard designated as IEC 62304 saw the light of day. This standard was also quickly harmonized according to the Medical Device Directive (93/42/EEC). The adoption of the then-new standard took some time and was not without its opponents. Manufacturers were perhaps already accustomed to requirements elicitation, for example, but the new standard raised the level of the ask via setting more expectations on what is to be covered in Clause 5.2.2. Another point of contention was traceability, which caused considerable head-scratching then and still remains a common point of failure in operations today. Risk management, too, became a much more involved process with the introduction of the requirements in Clause 7.

The IEC 62304 standard brought assessment organizations an entirely new assessment method, and one they then had to get to grips with in order to start using the standard. Prior to the standard, assessment of software had used IEC 60601-1 Clause 14, but due to the limitation of that standard to physical medical electrical equipment standalone software was effectively left out. This was also true of IEC 60601-1-4 which in practice addressed embedded software only. The IEC 60601-1 family of standards did represent an improvement over prior practices of assessing so-called single fault conditions where individual fault conditions were simulated to study how the device coped with or recovered from these faults. An early example here is given by IEC 60601-2-34 on invasive blood pressure measurement where testing required the use of a specific testing algorithm whereby an erroneous timer state was created in the software and the subsequent safety of the hardware was then assessed.

The standard was updated in 2015 when the most significant change was the addition of a pathway for legacy software, the so-called "Legacy Concept" (see Section 4.4). This established an alternative route to demonstrate safety and compliance in the case of products that had already been on the market for a long period of time.

The scope of the standard is discussed in detail in Section 3.2.3.

3.1.2 Modern Times

The IEC 62304 standard established a means to evaluate software even when that software was not a part of an electrical medical device. The standard was a much-needed addition to the toolbox of the medical-device industry, which had in fact for some time already utilized server-based software for the storage and analysis of data in healthcare applications. The new standard thus caught up with the development of technology.

After IEC 62304, a new much-needed pathway was in place to address software independently of a device, as depicted in Figure 3.3. The independence is not absolute, though, as the standard does remind the reader to also observe the host device somehow (see Section 3.2.3), but this is left for other standards to cover. Note how in the figure the paths of embedded software and software in its own right converge before compliance is demonstrated. Embedded software may be assessed as part of its host device, provided that IEC 62304 considerations are observed. Similarly, the assessment of standalone software must observe, in some way, what the host device is.

Today, when IEC 62304 is used in the context of assessment against the modern General Safety and Performance Requirements (GSPR, Annex I) of the EU Medical Device Regulation (EU 2017/745), a new step is needed. As technology, and the world too in general, have evolved since the last amendment to IEC 62304 in 2015 new requirements have been introduced for, e.g., validation (EU MDR GSPR 17.2), cybersecurity (EU MDR GSPR 17.4), usability (EU MDR GSPR 5) and the documentation accompanying a device (EU MDR GSPR 22.x). Subsequently, the adoption of another standard, IEC 82304-1, may be beneficial to the manufacturer.

The details of the IEC 82304-1 standard, too, are largely beyond our discussion here (although see Section 12.2). Nonetheless, this is a particularly practical standard for use with health software that is designed to operate on general computing platforms and intended to be placed on the market without dedicated hardware. In keeping with modern thinking, the standard also covers software validation, installation, maintenance, and disposal. It should

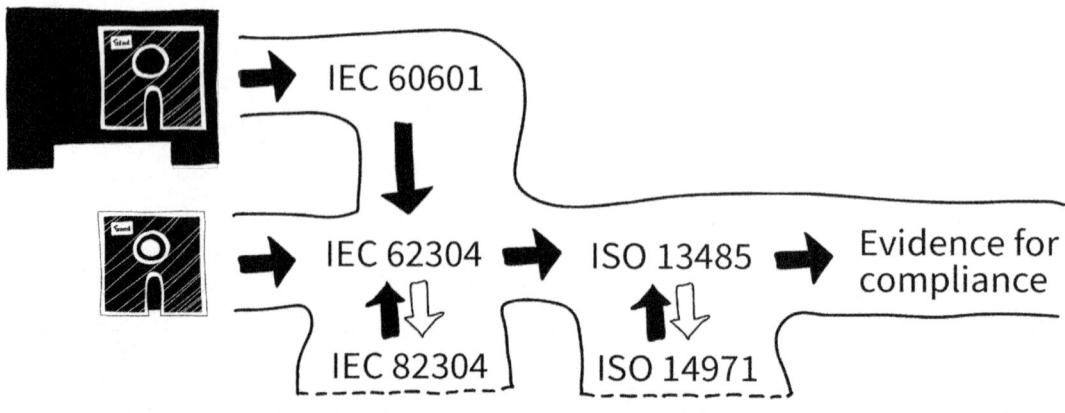

Figure 3.3 Demonstration of compliance with standards IEC 60601-1 Clause 14, IEC 62304, and IEC 82304-1.

be noted that the standard speaks of health software which is a term that may or may not be identical to medical-device software (see Section 12.2). At the very least, the use of the standard may help guide the development of health software and provide practical tips for the development of medical-device software.

The IEC 62304 continues to be adopted for use with medical-device software today. It provides a practical framework for structuring and analyzing the software-development work of manufacturers both within and without the organizations. As is one of the key intentions behind any standard, it enhances compatibility between expectations and compatibility between products that need to somehow link up whether during development or once available as finished products. It is not a perfect standard, though, and it is getting quite old.

3.1.3 The Future

The IEC 62304 standard provides a framework for the assessment of software safety and conformity to requirements. The elephant in the room, or the 800-pound gorilla cohabiting your small office cubicle, is that the latest amendment to the standard is from 2015 and by now woefully out of date.

It is possible to apply IEC 82304-1 (see above) as a patch to wrangle the IEC 62304 to the present day. This may be akin to what the Tyne Daly character attempts to do in the third installment of the *Dirty Harry* franchise – drag a Neanderthal to the 1970s – but it may be what you want to do.

This would, however, not remove all uncertainties. The following are some of the top concerns we hope are addressed when the IEC 62304 is finally revised:

- The requirements in the standard are written on a very high level that then leads to the possibility of significant differences in the acceptance criteria employed by different auditors. Better guidance and examples on the use of the standard in conformity assessments would thus be welcome and could be provided, e.g., as appendices to the standard or as a separate technical report.
- The handling of cybersecurity should be greatly improved. The current passing reference in Clause 5.2.2 during requirements analysis is not adequate or on par with modern medical-device regulation. The future version of the standard should address the requirements for both the implementation and the management of cybersecurity.

- The safe use of software will often be impacted by user interaction, user interfaces, and other usability considerations. The future version of the standard should thus also address these requirements in a practical way.
- Work on revising the standard has been bogged down in issues for the better part of its existence. One issue appears to have been the scope of the standard, which appears to be in limbo between strictly regulated medical devices and the more loosely defined health devices. It is our understanding that the standard is applicable to the assessment of software placed on the market as, or as part of, medical devices under the EU Medical Device Regulation. The uncertainty in the appropriate scope should be cleared up, perhaps by defining the scope as applying to medical devices and being optionally applicable to health devices.
- The standard relies heavily on an ABC software safety classification to levy requirements based on the criticality of the software under consideration (see Section 4.3). This is a good feature of the standard, but the assignment of the safety classification could do with more concrete guidance and examples on how each of the three classes is assigned. In other words, how the classification can be more accurately and uniformly applied and assessed by the users of the standard.
- Similar to the previous point, more guidance should also be provided on the requirement to rely only on risk-control measures external to the software (see Section 4.3). This requirement is not clear, particularly since a medical device may consist of several software systems each perhaps providing risk-control measures over the others.

In practice, we feel that the IEC 62304 standard should be accompanied by a technical report to provide the necessary guidance and examples needed to bring the standard to the present day. To make up for lost ground, we feel that a roadmap should be developed to gradually bring the standard in line with modern expectations both now and perhaps in the future too. This roadmap needs to address the increasing complexity of software, the management and implementation of cybersecurity, the use of various cloud-based architectures, and the evolving expectations of general safety and performance for software-based products. Some of these topics will also be covered later in Section 12 of this book.

In the next section we will dive into the contents of the present version of the standard, and also comment on the changes we feel are warranted.

3.2 Contents

The standard is roughly 40 pages of clauses and 47 pages of appendices in its original 2006 form with an additional fatly spaced 51 pages of changes made by the 2015 amendment. The appendices are there as guidance, to enhance your understanding and provide suggestions on the interpretation of the standard.

The business end of the standard consists of a foreword, an introduction, and the clauses themselves. The clauses are where the actual requirements of the standard are given, split into nine topics, and administered through the safety classification of your software and the safety classifications of the software items it consists of according to your design. Also, legacy software has its own path through the standard, as discussed later in Section 4.4.

The nine main clauses are broken down into numbered subclauses, which together give you the coordinate system often quoted when discussing the requirements imposed by the standard. We will go through each clause in later sections of this book, but the following provides the first bird's-eye overview of what to expect in the standard. This is the "lay of the land" you will be working with as you shape or reshape your own processes to comply with the standard as leanly and as smartly as possible. Note also that the numbering of Sections 4 through 9 of this book match with the clause numbers of the standard for easy reference.

3.2.1 Foreword

This is the foreword to the standard by the International Electrotechnical Committee (IEC), which has ownership of the IEC 62304 standard. The foreword provides an overview of how IEC standards are developed and expresses the goal of the standards to represent, as nearly as possible, an international consensus of opinion on any topic addressed.

The foreword explains that the IEC does not provide any marking mechanism to indicate its approval of, for example, devices manufactured according to its standards. This is why you don't get to apply an IEC 62304 certificate to your processes, but you may engage the services of several different types of consultants and certification bodies to check your product – and by extension your product-realization processes – for conformity with the standard. The outcome of such a check-up is a test report containing an assessment of how you meet the requirements of the standard, but not a definitive certificate per se. The foreword also forewarns you that no liability can be placed on the IEC or its staff. The conformity assessment and test report are discussed in more detail in Section 13.

The foreword then describes how the standard uses formatting to highlight requirements and definitions (printed using Roman type), informative notes (in small text), and terms (in small capitals) throughout the body of the standard. Finally, an asterisk in front of a section title indicates that Appendix B of the standard may be consulted for more guidance. The important detail to notice here is that any notes given by the standard are there in a non-binding informative role, as hopefully helpful comments that don't have the stature of requirements in the standard.

3.2.2 Introduction

This is the introduction that walks you into the world of the standard and gently sets up the scene for all that is to come. In so doing, the section is roughly equivalent to Clause 0 in ISO 13485. In both standards this introductory text comprises roughly 8–9% of the entire body of the standards and is more impactful than just thanking you for selecting this particular standard to read. In fact, the introduction packs a punch in setting up what you should expect from the standard, and how the writers of the standard see you making use of it.

The introduction begins by establishing software as an often-integral part of medical devices and then sets out to provide a framework of life-cycle processes to ensure the safe design and maintenance of such software. It grounds the assessment of software safety and effectiveness in the intentions for its use, and the evidence of those intentions being fulfilled without what could be considered unacceptable risks. Note that the terms "intended use", "intended purpose", and "clinical evidence" are not used here, but the implications are similar.

The description makes it clear that the standard assumes all work to take place within both a quality management system (covered in more detail in Clause 4.1; see Section 4.1) and a risk management system (covered in more detail in Clause 4.2; see Section 4.2). Within this context, the standard intends to define a four-process framework for the safe design and maintenance of medical-device software. The four processes presented are:

- "Development" (Clause 5; see Section 5)
- "Maintenance" (Clause 6; see Section 6)
- "Configuration management" (Clause 8; see Section 8)
- "Problem resolution" (Clause 9; see Section 9)

In case you are wondering what happens in Clause 7 which is missing from the above list, that is dedicated to risk management and is a topic we will address in Section 7. Note that despite introducing four distinct processes here, the standard still sets itself up overall as providing a framework of life-cycle processes for the safe design and maintenance of medical-device software – thereby subjugating configuration management and problem resolution to these grander goals.

The introduction further explains that all these processes are to be understood as consisting of activities, and the activities then consist of tasks. Each of these Lego blocks is given a definition in Clause 3 of the standard.

- **Process**

    Defined as a set of interrelated or interacting activities that transform inputs into outputs (Clause 3.14). The standard appears to expect that a process always consists of more than one activity – although it would be tough to argue this was a required characteristic. The term is sourced from ISO 9000, and thus both IEC 62304 and ISO 13485 use the same source for the definition of the term, but due to workflow reasons happen to refer to different versions of the same standard (2000 versus 2015). The difference does not appear significant.

- **Activity**

    Defined as a set of interrelated or interacting tasks (Clause 3.1). This may also address the use of resources, as suggested by a bafflingly placed note to Clause 3.14. The term is not sourced from outside the standard, but instead defined here from scratch. Interestingly, ISO 9000:2015 does also define the term and what is more puts a distinct project-management slant on it to capture the smallest identified object of work in a project. Thus, a minor point of contestation exists here between activity as understood in IEC 62304 (a set of tasks) and ISO 13485 (smallest packet of work in a project). In practice, the difference does not appear significant.

- **Task**

    Defined as a single piece of work (Clause 3.31). This is the practical atom of activity in a process or an activity as the standard sees it. The term is not defined in ISO 9000, and thus any use of it in ISO 13485 is according to some commonly accepted dictionary definition. The dictionary definitions, (e.g., a piece of work to be done or undertaken), match well with the definition adopted by IEC 62304.

As noted above, some minor differences may be observed in how the above three concepts are used by the two standards, the aging IEC 62304, and the ISO 13485. The differences are slight and do not prevent a practical arrangement to be negotiated between the two standards. If altercation was the goal, some contestation could be found in how the granularities of some terms work across the two standards, but as even atoms can be split multiple times in laboratory conditions this seems hardly practical. What matters is that you know how you want to use processes, activities, and tasks – and that you are consistent in their use.

Importantly, no specific model of processes, activities, or tasks is forced on the user of the standard, but instead the user is expected to map these elements to their own chosen life-cycle model. This also means that the user is expected to map the four key processes of the IEC 62304 standard to their own processes. This mapping will then be a major topic of discussion when comparing any two manufacturers and the processes they may have for software development – not to mention a major undercurrent of any conformity assessment based on the standard.

No organizational structure or division of work is imposed on the user, either. Instead, it is only expected that the processes, activities, and tasks defined by the standard are completed if conformity with it is claimed. ISO 13485 will add the necessary organizational overhead here (see Juuso 2022).

Similarly, the IEC 62304 standard requires that tasks are documented, but the shape of this documentation is left to the user. No names, formats, or explicit content is specified for the documentation to be produced. These are both significant design features of the standard and ones that you may miss if you skip learning about the introduction of the standard. In practice, you should aim to match the process and document names of the standard if you are starting fresh. This way retaining a mapping between your processes and documents, and those expected by the standard will be as easy as possible. However, you don't have to do so.

Importantly, the current version of IEC 62304 makes a normative (binding) reference to ISO 14971 risk management, which it thus requires everyone to use along with whatever further software-specific requirements it lays out for risk management (see Section 7). The intention here is to make some further requirements in the standard to address the nature of software and to ensure that the risk management for the software and any device using that software occurs in some synchronized manner. It is noteworthy that the reference both here and later in Clause 2 is made without specifying a version of the risk-management standard, and thus the reference is always to the latest version of the standard whatever that happens to be (e.g., ISO 14971:2019).

The introduction also clarifies the use of two essential concepts which show up repeatedly in the standard.

- **"Establish"**
  Clarified as defined, documented, and implemented. Note that in comparison to ISO 13485 (which defined "documented" as something that is established, implemented, and maintained) here the sentiment is the same, but the clarification does not explicitly call out maintenance.
- **"as appropriate"**
  Clarified as appropriate in the context of a process, activity, or a task unless excluded providing a documented rationale. Note that here the clarification is somewhat looser than in ISO 13485 which states that if a requirement is necessary for meeting product requirements, regulatory compliance, carrying out corrective actions, or managing risks, it can't be excluded. Similarly, the standard here contents itself with a documented rationale instead of an acceptable rationale, and limits the context to a process, an activity, or a task. The difference between the two standards is noticeable, but in practice hard to argue. The key takeaway is that something qualified with "as appropriate" is not something you can just ignore because it doesn't seem relevant.

Finally, the introduction adopts the common definitions of shall (it is required), should (it is recommended) and may (it is permissible) into the standard. These are defined as you would expect, but if you are not accustomed to these concepts in a professional setting, make sure you use them as defined here. Note that the term can, used to indicate a possibility by the ISO 13485, is not mentioned here. Figure 3.4 shows a word cloud of what to expect in the introduction.

3.2.3 Clause 1: Scope

Clause 1 represents 6% of the standard and sets the scope for its application. Clause 1.1 reiterates the intention of the standard to provide life-cycle requirements for medical-device software by defining a common framework for the life-cycle processes expected. The framework consists of a set of processes, activities, and tasks as already set out by the introduction.

74 ■ *Medical-Grade Software Development*

> /software ₃₀ /software is ₅ /medical ₆
> /medical device ₆ /device ₇ /the software
> ₁₄ /this standard ₁₈
> /standard ₂₂ /life ₅ /life cycle ₅ /cycle ₅
> /processes ₅ /activities ₈ /maintenance ₅
> /requirements ₅ /process ₁₉
> /management ₁₀ /risk ₁₁ /risk management ₈ /risk
> management process ₆ /management process ₇
> /clause ₅ /activity ₆ /compliance ₇ /compliance with ₆

Figure 3.4 Word cloud for the Introduction.

Clause 1.2 defines the intended field of application for the standard. The clause was rewritten completely for the 2015 amendment and now instructs that the standard applies to the development and maintenance of medical-device software regardless of:

- **Nature of the software:** The standard applies whether the software is a medical device itself or embedded in one. A non-binding note further suggests that requirements regarding software of unknown provenance (SOUP) also apply and offers Clause 8.1.2 as a point of further reading. Note though that an entire medical-device software system in itself can't be a SOUP (see a non-binding note to Clause 3.29)
- **Mode of execution:** The standard applies whether the software is executed on a processor directly or via an interpreter
- **Type of storage:** The standard applies regardless of the type of persistent storage used for the software (e.g., flash memory, hard disk, optical disc)
- **Type of delivery:** The standard applies regardless of the method of delivery used for the software (e.g., via e-mail, optical disc, flash memory). Note that this presumably also covers cloud solutions, and the method itself is not considered to be a medical device by the standard

Clause 1.2 also points out that the standard does not cover software validation or final release, not even in cases where the software is a medical device itself. A non-binding note to the standard further elaborates that validation and other system-level development activities may be found in product standards such as IEC 60601-1 and IEC 82304-1 (see Section 12.2).

Clause 1.3 introduces the requirement that the standard is applied together with other relevant standards when developing a medical device. The list of relevant standards, and their mapping to this standard, is covered in detail in Appendix C of the standard, but this should not be considered exhaustive or up to date.

Clause 1.4 discusses how compliance with this standard may be claimed. Compliance is defined as the implementation of the processes, activities, and tasks defined in the standard as these are applicable to your software safety class. The essential concept of a software safety class is thus first introduced by this clause in passing, but with the momentous effect that not all the requirements of this standard need to be fulfilled if your software safety class is not implicated by the requirement. The concept itself will be covered a little later in Clause 4.3.

The evaluation of compliance is determined via two actions:

- Inspection of your documentation (incl. your risk management file)
- Assessment of your processes, activities, and tasks

A non-binding note to the clause further clarifies that the assessment may be carried out by internal or external audit, although the use of the former option has in practice been a contested issue during regulatory assessments of the final medical device in the EU. It should also be noted that the standard appears to use the term inspection for the first action, the term assessment for the second action, and the term determination for the overall evaluation of compliance. This distinction here is less-than-optimal as taken together with the note on "assessment" it appears to suggest that only the second action could be performed internally or externally. In practice the whole IEC 62304 evaluation may be performed internally or externally, but the outcome is still only as valuable as the body conducting the final regulatory assessment of the medical device deems it. Using a reputable certification body is recommended here. See Section 13 for more discussion on conformity assessment.

Further non-binding notes to the clause acknowledge the need for flexibility in assessing the methods of process implementation and the performing of activities and tasks and call for documented justification in cases where a requirement marked as as appropriate by the standard is not fulfilled. Finally, a non-binding note offers Clause 4.4 (see Section 4.4) as a

Figure 3.5 Word cloud for Clause 1.

further reference in the case of legacy software, and another non-binding note comments that conformance may mean the same as compliance (e.g., in the case of ISO/IEC 12207).

The whole of Clause 1.4 is somewhat unusual in the sense that ISO 13485, for example, does not explicitly instruct how compliance is to be addressed. The clause may be dropped from future editions of the standard, but some mechanism of placing stricter requirements on software associated with a higher level of concern (e.g., software for vital signs monitoring) as opposed to a lower level of concern (e.g., mobile applications for personal sleep monitoring) will no doubt endure. Figure 3.5 shows a word cloud of what to expect in the clause.

3.2.4 Clause 2: Normative References

Clause 2 (less than 1% of the standard) makes one normative (binding) reference to the ISO 14971 standard on the application of risk management to medical devices. Note that the reference is undated implying that the latest version of the standard is implicated. In the case of ISO 14971 this means a version of the risk-management standard that is a decade newer than the version available when the IEC 62304 standard was amended in 2015. In any case, the ISO 14971 standard is to be understood as indispensable and explicitly required for the application of IEC 62304 for the time being.

A later edition of IEC 62304 may remove the normative reference to ISO 14971 as new types of software, perhaps also including mobile applications and other health software, appear alongside traditional full-on

/references³

Figure 3.6 Word cloud for Clause 2.

medical-device software. The requirement to apply hardcore ISO 14971 risk management to all types of software has been one of the many points of debate for the next revision of IEC 62304. We see ISO 14971 as the recommended approach to risk management in the context of medical devices but recognize the need to enable other lower-concern users to begin adopting elements of the IEC 62304 without the immediate issue of implementing full ISO 14971 risk management. Note also that in ISO 13485 a mention of a risk-based approach does not automatically always refer to ISO 14971 risk management (see Juuso 2022). Figure 3.6 shows a word cloud of what to expect in the clause – this time the view is not only easy to interpret but also quite conclusive.

3.2.5 Clause 3: Terms and Definitions

Clause 3 comprises approximately 16% of the standard, all of which is real estate it uses to define important terms used throughout the rest of the standard. The amended version of the standard introduces 39 terms, out of which five terms are new to the 2015 edition. One term (software product) was rescinded in the amendment. In total, 12 terms are defined from scratch while the remaining 27 terms are based on definitions made elsewhere (e.g., ISO 14971, ISO 13485, IEC 60601-1, ISO 9000, ISO/IEC 12207). Out of all the terms now defined, the following group of four terms, presented from top to bottom, is perhaps the most essential set of concepts for understanding the clauses of the standard.

- **System**
  An integrated composite (collection) of processes, hardware, software, facilities, and people that acts to satisfy a need or objective (Clause 3.30).
- **Software system**
  An integrated collection of software items organized to accomplish a function or set of functions (Clause 3.27). This is the highest of three levels decomposing software. Note that medical device software (Clause 3.12) is then defined as a software system developed for the purpose of

a) being incorporated into a medical device or b) intended to be used as a medical device.

- **Software item (incl. SOUP)**

    An identifiable part of a computer program (Clause 3.29). This is the midlevel of three levels decomposing software. On a related note, if that software item is a) developed elsewhere for a purpose other than being incorporated into the software discussed here, and it is generally available, or is b) developed previously and adequate records of the development are not available, then it is considered software of unknown provenance (SOUP). Note that the entire software system can't be claimed as SOUP.

- **Software unit**

    The lowest level of software which is not subdivided into further parts. This is the lowest of the three levels of decomposing software. See Clause 3.28.

These four terms thus comprise four nested levels for software decomposition as the standard understands it, or the "Russian doll" of software and where it is used. Note that in a somewhat unhelpful move, the 2015 amendment offers source code, object code, control code, control data, or a collection of these as an example of a computer program, but does so in the definition of a software item. It should be emphasized that the examples listed apply equally as the representations of an identifiable part of a computer program whether that is a software unit, a software item, or a software system. They are not examples of a software item per se. In an equally unhelpful manner each of these – software unit, software item, and software system – may, according to the standard, also be called a "software item" by themselves. This chain of definitions is a veritable Gordian knot, an intractable problem much like the explanation behind the acronym PHP which is short for PHP Hypertext Pre-processor. Someone out there is laughing their head off, no doubt, but for a beginner trying to grasp IEC 62304 it suffices to remember that units build up items, items build software systems, and these are used in a wider system also including people. Note that the granularity here is left up to the manufacturer to define (note to Clause 3.28).

Also apparent from reading the list of terms defined is that the standard sees itself as extending ISO 14971 risk management for the special application area of software. It is thus not surprising that a total of 13 terms have been imported from the 2007 version of the ISO 14971 standard. These terms are: "harm", "hazard", "hazardous situation", "manufacturer", "residual risk", "risk", "risk analysis", "risk control", "risk estimation", "risk management", "risk management file" and "safety". The sourcing of these terms from the risk-management standard is a welcome choice promoting compatibility between the standards, and boding well for the adoption of the later ISO 14971 editions.

By comparison only one term, "medical device", has been imported from the 2003 version of ISO 13485. The definition of what is and what is not a medical device is, of course, a far bigger question than just a term definition and is something better left to the regulations in each jurisdiction. A further seven terms – "activity", "process", "release", "risk analysis", "system", "traceability", and "verification" – are silently inherited by ISO 13485 from ISO 9000:2015 and shared by the two standards here. Figure 3.7 shows a word cloud of what to expect in the clause.

3.2.6 Clause 4: General Requirements

Clause 4 (11% of the standard) addresses four fundamental topics: the quality management system (Clause 4.1), risk management (Clause 4.2), software safety classification (Clause 4.3) and legacy software (Clause 4.4). The aim of the discussion here is firstly to set up the foundations for developing software by building on quality and risk management, and secondly to set you off in the right direction for using the standard by instructing both the software safety classification, used in interpreting the standard, and the special

Figure 3.7 Word cloud for Clause 3.

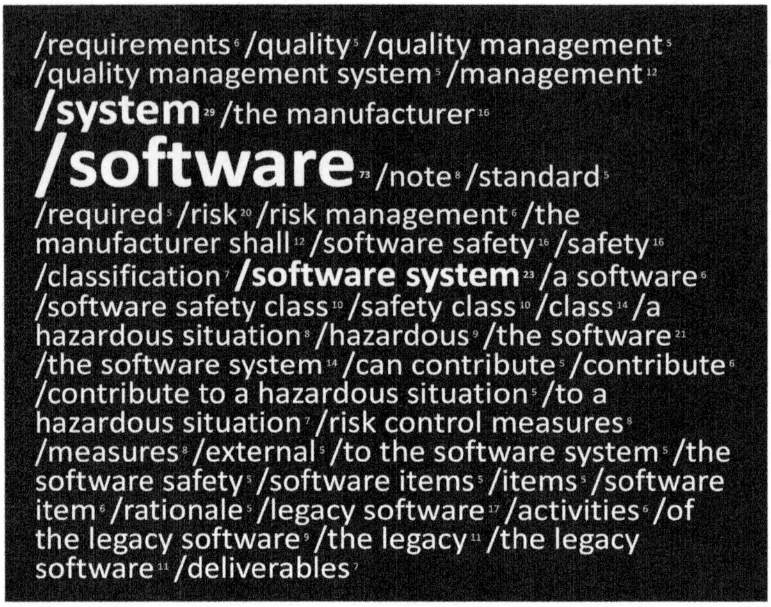

Figure 3.8 Word cloud for Clause 4.

pathway for legacy software, should that be your situation. The clause and its subclauses will be covered in detail in Section 4 of this book. Figure 3.8 shows a word cloud of what to expect in the clause.

3.2.7 Clause 5: Software Development Process

Clause 5 is the heart of the standard. It comprises roughly 36% of the whole standard and dives deep into how your software development is to unfold. This is the clause that speaks to your software-development life cycle, and that must match with your software-development process. The clause covers development planning, requirements analysis, architectural design, detailed design, unit implementation and verification, integration and integration testing, software-system testing, and release. The clause and its subclauses will be covered in detail in Section 5 of this book. Figure 3.9 shows a word cloud of what to expect in the clause.

3.2.8 Clause 6: Software Maintenance Process

Clause 6 is the yin to the yang represented by the preceding clause of software development. In other words, this is the Nick Nolte to your 48-hour tour of the town with Eddie Murphy, or what you need to do after all the

> **/software** 201 /software development 28 /development 38 /process 11 /planning 7 /software development plan 16 /development plan 17 /plan 32 /the manufacturer 57 /the manufacturer shall 55 /the manufacturer shall establish 5 /manufacturer 57 /shall establish 5 /establish 7 /a software 5 /activities 18 /of the software 11 /**the software** 61 /the software development 18 /the software system 8 /software system 25 /system 40 /life 7 /life cycle 6 /processes 5 /note 43 /deliverables 8 /documentation 5 /activities and tasks 8 /tasks 10 /system requirements 11 /requirements 56 /software requirements 20 /test 22 /risk 17 /risk control 10 /risk control measures 5 /control 15 /control measures 6 /implemented 6 /software configuration 7 /configuration 13 /management 13 /including 10 /soup 11 /items 29 /support 6 /software problem resolution 5 /problem 7 /problem resolution 6 /the medical device 12 /the medical device software 9 /medical 15 /medical device 15 /medical device software 11 /device 18 /device software 11 /class 56 /class a 22 /software items 18 /item 9 /integrated 9 /the software development plan 11 /reference 12 /design 13 /in the software 10 /in the software development plan 9 /the manufacturer shall include 11 /the manufacturer shall include or reference 8 /shall include 13 /shall include or reference 9 /include 22 /procedures 13 /the system 5 /integration 24 /verification 20 /between software 7 /tools 5 /the manufacturer shall include or reference in the software development plan 6 /class c 7 /software integration 8 /integration testing 10 /testing 24 /the software items 5 /perform 7 /class b 27 /it is acceptable 5 /acceptable 6 /software system testing 9 /set 5 /information 8 /required 6 /acceptance 8 /acceptance criteria 7 /criteria 10 /see clause 5 /document 20 /software configuration management 5 /maintenance 5 /problem resolution process 5 /develop 5 /examples 11 /defects 5 /relevant 5 /analysis 6 /examples include 7 /hardware 8 /unit 15 /interfaces 9 /interfaces between 5 /related to 5 /evaluate 7 /ensure 10 /verify 11 /the manufacturer shall verify 5 /implement 5 /tests 11 /architecture 14 /segregation 6 /software architecture 6 /detailed 7 /detailed design 6 /software units 10 /units 10 /each software unit 5 /software unit 13 /the manufacturer shall document 6 /verification is 7 /the test 7 /results 7 /the integration 9 /anomalies 8 /released 7

Figure 3.9 Word cloud for Clause 5.

frolicking of developing a piece of software. The clause comprises roughly 6% of the entire standard and covers topics such as maintenance planning, problem and modification analysis, and modification implementation. The clause may be easy to overlook during the initial excitement of software development, but you will need to address these issues before you get an

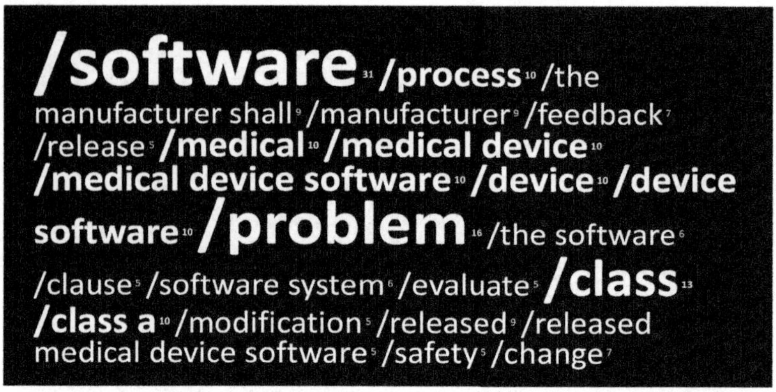

/software 31 **/process** 10 /the
manufacturer shall 9 /manufacturer 9 /feedback 7
/release 5 /medical 10 /medical device 10
/medical device software 10 /device 10 /device
software 10 **/problem** 16 /the software 6
/clause 5 /software system 6 /evaluate 5 **/class** 13
/class a 10 /modification 5 /released 9 /released
medical device software 5 /safety 9 /change 7

Figure 3.10 Word cloud for Clause 6.

approving nod from your notified body, should you need it, and before you release your product. The clause and its subclauses will be covered in detail in Section 6 of this book. Figure 3.10 shows a word cloud of what to expect in the clause.

3.2.9 Clause 7: Software Risk-Management Process

Clause 7 dives deep into the question of how ISO 14971 risk management should take place in the context of software medical devices. The clause comprises approximately 7% of the standard and covers topics such as analysis of software contributing to hazardous situations, risk-control measures and their verification, and risk management of software changes. After a series of fairly sequential clauses from 1 to 6, Clause 7 is the first clause that folds into the previous discussion and wants to see risks managed throughout the life cycle in a helpful way. The clause and its subclauses will be covered in detail in Section 7 of this book. Figure 3.11 shows a word cloud of what to expect in the clause.

3.2.10 Clause 8: Software Configuration Management Process

Clause 8 (approx. 5% of the standard) has the important task of introducing the concept of configurations (e.g., version-identified pieces of software used in your product, and the version-identified product itself) and changes to them. The requirements here look at your overall approach to configurations, changes to configurations (incl. the software behind the version number), and retaining an adequate history for the controlled items. The

Figure 3.11 Word cloud for Clause 7.

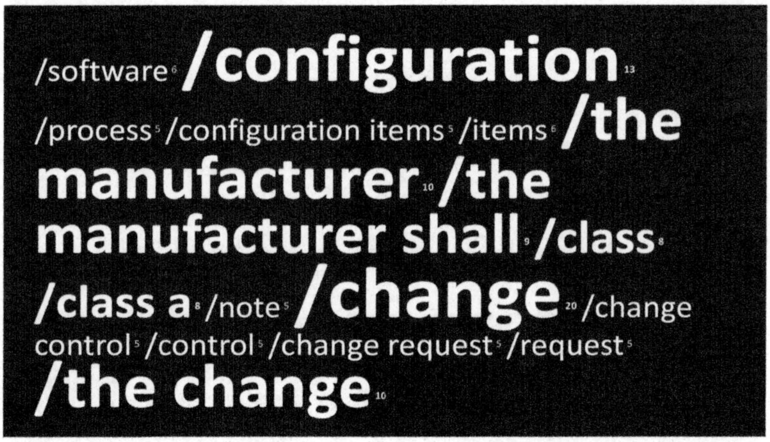

Figure 3.12 Word cloud for Clause 8.

clause and its subclauses will be covered in detail in Section 8 of this book. Figure 3.12 shows a word cloud of what to expect in the clause.

3.2.11 Clause 9: Software Problem Resolution Process

Clause 9 (approx. 5% of the standard) has the honor of closing out the standard by instructing problem resolution regarding your medical-device software. The clause covers topics such as problem reports, problem investigations, advising relevant parties, using your change-control process (from Clause 8), maintaining adequate records, analyzing problems for trends, verifying problem resolutions, and test documentation. The clause and its subclauses will be covered in detail in Section 9 of this book. Figure 3.13 shows a word cloud of what to expect in the clause.

3.2.12 Annex A – Rationale for the Requirements of This Standard

The first of four informative-only annexes, Annex A, provides a brief rationale for the existence of the standard as a means of imposing an improved level of control over the development of safe medical software. The notion here is that the mere testing of software is not adequate to ensure its safety in operation. Instead, the requirements of the standard are to be met in designing the processes for software development and maintenance as is appropriate for the risks associated with the type of software in question. Adherence to these processes is even positioned as the primary requirement of the entire standard by this annex. As prime directives go, this one really should not catch you off guard like that secret fourth directive did Alex Murphy towards the end of *Robocop*.

The 2015 amendment of the standard only made minor changes to the annex. In addition to an updated table providing a non-normative overview of which requirements map to which software safety classes, the new text made four fairly cosmetic changes to the text. The changes made are not worth commenting on here in any great detail, but if you find yourself in a

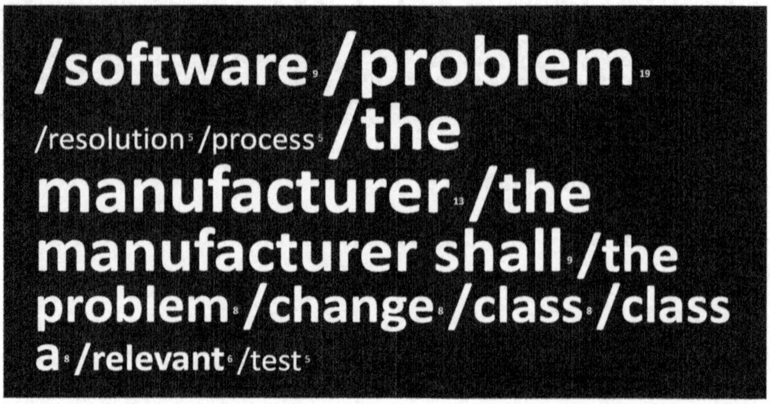

Figure 3.13 Word cloud for Clause 9.

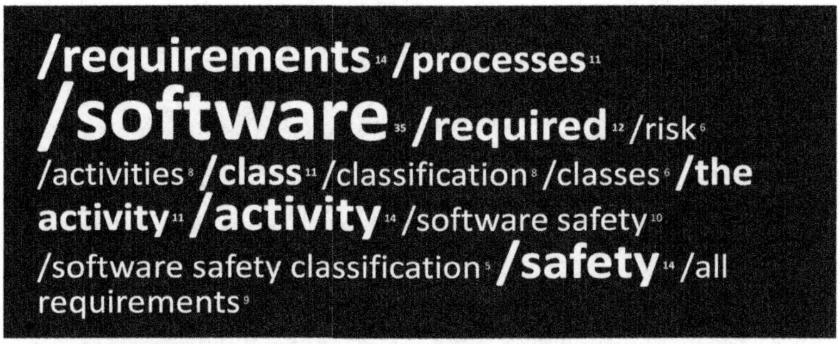

> /requirements₁₄ /processes₁₁
> /**software**₃₅ /required₁₂ /risk₆
> /activities₈ /**class**₁₁ /classification₈ /classes₆ /the
> activity₁₁ /**activity**₁₄ /software safety₁₀
> /software safety classification₃ /**safety**₁₄ /all
> requirements₉

Figure 3.14 Word cloud for Annex A.

heated argument over whether software can contribute to hazards (old text) or hazardous situations (new text), or whether the standard thinks it is easy (old text) or just possible (new text) to assure class A software safety through overall medical-device design activities and life-cycle controls, check here for any discrepancies between you and your jousting partner. It is, however, worth noting that the annex, too, expects such hazards to be a part of your overall risk analysis activities.

Figure 3.14 shows a word cloud of what to expect in the annex.

3.2.13 Annex B – Guidance on the Provisions of This Standard

Annex B offers roughly 20 pages of commentary on the clauses of the standard.

Figure 3.14 shows a word cloud of what to expect in the annex. The commentary is not normative, as it does not place requirements on the adoption of the standard, but it does speak to the intentions of its writers which may or may not correspond with the views of medical device regulators.

The 2015 amendment of the standard made mostly terminological changes to the annex (e.g., changing software product to "medical device software", "hazard" to "hazardous situation" in many places) and introduced two pages of discussion on legacy devices. The contents of the annex are discussed later in this book as needed, but for now, Figure 3.15 provides a succinct overview of the topics discussed in the annex.

3.2.14 Annex C – Relationship to Other Standards

Annex C plots the place of the IEC 62304 in the 2015 world of standards. Here the foundation for developing medical software is laid out as consisting

86 ■ *Medical-Grade Software Development*

/this standard[31] /standard[35] /provide[7] /development[46] /development process[12] /**process**[75] /quality[10] /safe[6] /**medical**[57] /**medical device**[53] /medical device software[29] /device[63] /device software[29]
/**software**[141] /activities[16] /tasks[12] /**the software**[105] /requirements[65] /change[20] /processes[18] /software development[14] /planning[7] /software requirements[20] /requirements analysis[7] /analysis[19] /architectural[7] /design[34] /detailed[17] /detailed design[11] /software unit[14] /implementation[14] /verification[6] /software integration[7] /integration[13] /integration testing[7] /**testing**[41] /software system[25] /software system testing[6] /**system**[61] /release[5] /life[14] /life cycle[5] /model[7] /dependencies[5] /inputs[10] /software safety[17] /software safety classification[7] /safety[31] /**risk**[72] /risk analysis[5] /established[7] /failure[7] /waterfall[5] /life cycles[5] /cycles[6] /strategy[6] /determine[7] /customer needs[5] /**define**[10] /implement[12] /test[12] /incremental[5] /system requirements[7] /part of[11] /complete[7] /evolutionary[4] /product[7] /outputs[11] /specifications[10] /documents[6] /approved[5] /**software item**[30] /software architecture[14] /architecture[23] /output[7] /configuration[17] /configuration management[7] /ensure[20] /state[5] /released[5] /development and maintenance[5] /maintenance[24] /**manufacturer**[12] /risk management[35] /component[7] /responsibility[5] /software risk management[7] /risk management process[13] /as part of[5] /software maintenance[10] /technical[5] /**problem**[28] /function[7] /modification[13] /requests[5] /regulatory[5] /quality management[7] /quality management system[5] /standards[5] /level[11] /software configuration management[7] /**software items**[35] /**items**[52] /software units[7] /methods[5] /software engineering[7] /establish[5] /intended[10] /identify[5] /identified[11] /scheme[5] /severity[7] /hazard[7] /risk control[20] /**control**[33] /interfaces[5] /hardware[7] /the architecture[5] /the software architecture[5] /**activity**[30] /expected[5] /correctly[6] /faults[5] /**class**[14] /class c[8] /software safety class[7] /requirement[7] /document[10] /plan[26] /procedures[10] /plans[7] /detail[6] /information[5] /verify[13] /behaviour[5] /documented[10] /functional[7] /data[8] /components[7] /affect[5] /specification[6] /subclause[5] /code[8] /hazardous[19] /coding[5] /documentation[5] /effect[5] /white box[7] /white box testing[6] /box testing[10] /maintenance process[10] /problems[11] /the problem[6] /a problem[5] /problem resolution[11] /software problem resolution process[7] /change requests[6] /hazardous situations[7]

Figure 3.15 Word cloud for Annex B.

of the big management standards such as ISO 13485 for quality management and ISO 14971 for risk management. On top of these you are expected to employ safety standards as is appropriate for the type of devices you manufacture. Chief among the usual suspects here is IEC 60601-1 for hardware and IEC 62304 for software. This much has not changed in the decade since the standard was published. A good number of other potentially relevant standards have been published (e.g., see Section 12), and newer revised versions of many standards have been published since, but the basic concept of fitting standards together and the biggest of the building blocks have remained the same.

/relationship 17 /relationship to 14 /standards 10 /general 6 /this standard 55 /standard 62 /development 58 /maintenance 18 /medical 23 /medical device 13 /device 13 **/software** 211 /subsystem 16 /the medical device 5 /management 47 /annex 6 /provide 6 /develop 7 /safety 20 /specific 6 /medical devices 6 /part of 6 /detailed 15 /required 13 /methods 6 /tools 5 /implement 6 **/requirements** 90 /shows 5 /defined 9 /items 6 /requires 5 /manufacturer 7 /system 40 /clause 17 /related 10 /software development 25 /software development planning 5 /planning 20 /design 56 /design and development 10 /software requirements 17 /software requirements analysis 7 /analysis 24 /architectural design 13 /detailed design 14 /software unit 12 /unit 15 /implementation 47 /verification 56 /software integration 16 /integration 31 /testing 28 /software system 8 /release 10 /validation 33 /establish 5 /software maintenance 9 /plan 29 /control 38 /problem 31 /modification 22 /modification analysis 6 /modification implementation 9 /hazardous 5 **/risk** 68 /risk control 21 /risk control measures 12 /risk management 31 **/configuration** 26 /configuration identification 7 /identification 13 /traceability 5 /change 9 /software problem resolution 8 /problem resolution 19 **/process** 80 /risk management process 10 /hazards 5 /software safety 6 /classification 5 /evaluation 5 /verify 9 /document 11 **/pems** 84 /pems requirements 14 /requirements for software 5 /electrical 5 /include 11 /addressed 5 /pems development 9 /model 5 /component 6 /level 12 /specification 15 /pess 7 /architecture 23 /unit verification 8 /pems validation 20 /requirement 11 /pems validation plan 5 /validation plan 6 /test 12 /software subsystem 6 /results 13 /activities 16 /the risk management file 5 /development process 5 /software development plan 10 /require 9 /life 16 /life cycle 9 /cycle 16 /development life cycle 5 /maintenance process 8 /processes 24 /additional 6 /identified 7 /configuration management 7 /documentation 14 /subsystem of a pems 5 /essential performance 6 /performance 6 /application 6 /failure 6 /documents 7 /activity 22 /reference 5 /software validation 7 /system level 7 /scope 7 /risk control measure 5 /control measure 5 /pems development life 6 /documented 10 /the verification 6 /software verification 5 /criteria 7 /identify 9 /issue 5 /network 16 /data 19 /data coupling 10 /subsystems 6 /the network 9 /the design 5 /independence 5 /scope of this standard 5 /responsible 6 /equipment 13 /evaluate 6 /related requirements 5 /qualification 12 /ivd 6 /process implementation 10 /activity task 10 /task 10 /software coding and testing 6 /software qualification testing 10 /release management and delivery 5 /configuration control 5

Figure 3.16 Word cloud for Annex C.

Figure 3.17 Word cloud for Annex D.

The 2015 amendment updated the references from 2006 but did not alter the approach to using IEC 62304 in any significant way in Annex C. Figure 3.16 provides a succinct overview of the topics discussed in the annex.

3.2.15 Annex D – Implementation

Annex D is a brief two-page dip into how manufacturers can adopt the IEC 62304 standard in practice. The gist of the discussion is that medical-software development is required to take place within the context of a quality management system (see Section 4.1). This QMS does not, however, always necessarily need to be certified in order for this standard to be adopted. If a QMS or even a subset of it in the form of some relevant processes (e.g., regulation of development, verification, validation) exists before the adoption of the standard, a gap analysis of some sort should be performed to see how that QMS should be revised or whether the required processes should be separately described. Maintaining separate processes may get to be more trouble than it's worth in the long term so our advice is to read this book, and then marry the quality management system and software-development requirements together in a practical way.

The annex also notes, in full agreement with ISO 13485, that the manufacturer retains responsibility for any outsourced software development. Note though that the annex serves this sobering assertion with the provision that the suppliers do not have a documented QMS of their own in place. ISO 13485 is not so lenient.

Finally, the annex offers up a sketch of a checklist for small manufacturers without a certified QMS. This is, in practice, a table of contents for the standard with a checkbox and space for defining actions for following the manufacturer's own assessment of how well the requirements of the clause are met at the moment. The annex was not updated by the 2015 amendment. Figure 3.17 provides a succinct overview of the topics discussed in the annex.

Chapter 4

GENERAL REQUIREMENTS

This is where you are still infatuated with the new-car smell of your fancy new ride but are silently slapped with a hefty insurance bill for the privilege of actually driving it. Or in cinematic terms, you are riding high after a dream wedding oblivious to the fact that your spouse is a covert operative for some world government. In any case, "general requirements" is a soft way of saying heads up this will hurt. This is where you get to learn about the heavy lifting awaiting you before that Schwarzenegger physique is looking at you from the mirror. The good news is that you may already have a QMS and a risk-management system in place, but even then, this clause may have news for you regarding the software safety classification and handling of legacy software.

General requirements are the topic for the entirety of Clause 4 in IEC 62304. The clause comprises a deceptively slender 4% sliver of the standard. The four main topics under discussion are:

- Quality management system (Clause 4.1)
- Risk management (Clause 4.2)
- Software safety classification (Clause 4.3)
- Legacy software (Clause 4.4)

Clause 4 could just as well be called "prerequisites", "foundational requirements", or "the forever plan to the perfect software pipeline", but it is not. Whatever the title of the clause, the amount of heavy lifting to meet its requirements is formidable. The good news here is that your QMS will already meet much of what is required. In other words, your quality- and risk-management processes, as well as their instructed outputs, will form an essential part of the documentation assessed in any IEC 62304 or ISO 13485 conformity assessment (see Section 11).

In the following, we will address each of the four subclauses in detail. Note that the numbering of the following subsections corresponds to the clause numbers in IEC 62304.

4.1 Quality Management Systems

Clause 4.1 of IEC 62304 requires that the manufacturer is able to demonstrate their ability to consistently meet both customer and applicable regulatory requirements. The implied requirement to implement a quality management system (QMS) to do so is stated in the heading of the subclause and in the non-binding notes to the clause. Quality management and quality management systems (QMS) are the central topics of the ISO 13485 standard, and in practice best left for a discussion of that standard (see Section 2).

The IEC 62304 standard, much like the ISO 13485 standard too, leaves the door open to adopting other alternative national standards and regulations on quality management systems. How you obtain the evidence for the required demonstration is largely up to you, but you must have the evidence and it should be based on the QMS you run. In practice, choosing something other than ISO 13485 as your bridgehead into QMS here will be a Dunkirk waiting to happen, or a dodoist exercise in escapism. Don't do it unless you well and truly know the desert island you want to inhabit.

A further note on the clause offers ISO/IEC 90003 on guidelines for the application of ISO 9001 to computer software as further reading on the intersection between quality management and software. Note, however, that the standard in question has undergone several revisions since the publishing of the IEC 62304 standard and that ISO 9001 too has been revised since the ISO 13485 standard was forked from it. If your intention is to get certified to ISO 13485 you should not lose sight of that goal here.

In practice, implementing the IEC 62304 standard as part of your organization's overall QMS will lead to some software-specific requirements and a process or two described as standard operating procedures (SOPs) in your QMS (see Section 10). These requirements will be discussed in the coming sections of this book.

4.2 Risk Management

The topic of risk management is addressed in Clause 4.2 of the IEC 62304 standard swiftly and mercilessly: you must adopt ISO 14971 risk management. As elsewhere in the standard, except for the term definitions in Clause 3, the reference to ISO 14971 is made without specifying a version of the standard which is to be interpreted as meaning the latest version of the standard published today. This requirement, and indeed the whole clause itself, is only 12 words long, but the requirement is a profound one – at least for now.

In later as-of-yet unpublished revisions of the IEC 62304 attempts have been made to rework this clause in a way that is compatible with ISO 14971 but does not necessarily require the adoption of that standard. If these changes came about, the general relationship between the two standards would remain, but lower-risk health or wellness software could more readily adopt IEC 62304 life-cycle processes without the need to also adopt ISO 14971 risk management. This would hopefully bring welcome rigor to the development of non-medical software, and bring them within a more feasible jumping distance if they then someday need to make the transition to medical software.

The issue of risk management has been widely reported as one of the biggest flashpoints in revising the IEC 62304 standard in the past years as some experts see it as an absolute must and others as too much to ask of every class of software. Time will tell where the chips fall on this matter, but in general, adopting one standard should not really require the adoption of also another standard – you claim conformity with the standards you meet and not the standards those standards may link to in one way or another. If you wish to adopt both IEC 62304 and ISO 14971 you may of course always do so, and in general, this is likely to be a smart move, but bundling the two standards and asking manufacturers to take it or leave it seems like a bridge too far.

In practice, implementing the IEC 62304 standard as part of your organization's overall QMS and risk-management activities will lead to some software-specific requirements in your product-related risk-management activities. These requirements may affect your standard operating procedures for risk management and product realization, SOP-8 and SOP-10 as discussed in our QMS book (Juuso 2022).

As a final practical note on reacting to new versions of any standard, you will want to apply consideration and restraint in adopting any new version.

Adopting a new version is usually not just a copy-replace operation in MS Word. You will want to ensure you know what has changed in the new version, how this affects your operations, and how you will go about implementing the necessary changes before you contemplate pulling the trigger. You won't have all the time in the world to ponder the change but don't jump in blindfolded either.

In the IEC 62304 standard, product-specific risk management is addressed in greater detail in Clause 7 (see Section 7). Clause 7, as you will later see, makes heavy use of something called a software safety classification (explained in the next section) to attach more requirements on higher-risk software and allow lower-risk software to be developed more speedily. It is, however, important to note that Clause 4.2 here expects to see risk management take place according to ISO 14971 for all classes of software regardless of their software safety classification.

4.3 Software Safety Classification

Clause 4.3 of the standard is dedicated to the essential topic of software safety classification, or how to address different levels of concern in software. Think of this classification as setting up the fault lines in the tectonic plates that make up the standard and as your key to finding the right combination of activities that unlocks the vault of medical software development for you.

The ABC classification described here will cut through the implementation of Clauses 5 through 9 of the standard based on the risk profile of the software you are developing. The idea here is much the same as in the risk-based approach employed in ISO 13485 quality management: resources are allocated to, and more scrutiny is placed on, areas where risks are greater. As a result, low-risk software and software components may be classified as belonging to class A and thus dealt with in a simpler fashion than their high-risk counterparts which are classified into class C.

The formula for assigning the safety classification is quite straightforward but figuring out the harms and risks associated with the device, and tracking these down to individual software items sooner or later, will require effort. In the following, we will first look at the basic formula for assigning software safety classifications and then look at some fundamental concepts and interpretations that feed into arriving at the correct safety classification. Later in Section 7 we will also look at how some of these same stipulations bleed over to the actual risk management of your device.

4.3.1 Classification Formula

The 2015 amendment offers a helpful workflow to assign a safety classification to software or a software component. This workflow can be summarized as the following four rules:

- **All software is class C unless otherwise successfully argued.** In other words, you can't, for example, forget to perform the classification or lose documentation and say that the software was something else than C.
- **If software failure can't lead to a hazardous situation, or if that failure doesn't result in unacceptable risk after mitigation the software is class A.** In the latter case, the risk-control measures (mitigations) are to be external to, or independent from, the software in question, and you must evaluate their effectiveness. The definition of acceptable risk is for you to argue beforehand and in conformance to ISO 14971 (see that standard's Clause 4.2 on management responsibility in particular).
- **Otherwise, if the severity of the possible harm is non-serious the class is B.**
- **Otherwise, if the severity of the possible harm is serious the class is C.**

The above rules may be condensed into the following simple decision diagram (see Figure 4.1) that leads to the assignment of the ABC safety classification after answering a maximum of three questions. Remember that until these questions are answered the classification is C, that all software may be expected to fail eventually, and that only external mitigation (risk-control measures) may be considered here. Also remember that, as explained below, any child components will inherit the safety class of their parent unless otherwise successfully argued.

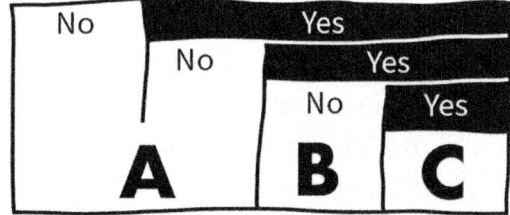

Figure 4.1 Key to safety classification.

In practice, when assigning a software safety classification, the manufacturer must consider the possible ways the operation of the software may become flawed, what failures may be caused by its design, and also the results of the risk-management activities concerning the product. Note that here harm is only considered from the point of view of the patient, user, and other humans (unlike what is defined for harm in ISO 14971; see below). As noted above, the default software safety classification is C (corresponding to the highest risk) until a lower classification has been assigned based on risk-management activities and a valid rationale. To use the above classification formula, we will need to discuss the definition of harm and the sources of harm. This is addressed next.

4.3.2 What Is Harm?

It is noteworthy that when assigning a software safety classification, the present version of IEC 62304 only considers risk from the point of view of harm to the patient, operator, or other people (Clause 4.3a). Harm to equipment or the environment is not part of the prescribed picture here even though these are a part of the definition of harm borrowed from ISO 14971 in Clause 3.8. In this setting, harm is considered as resulting from a hazardous situation the software can contribute to in a worst-case scenario. The implication here is that the software safety classification is based on something less than the full ISO 14971 risk matrix you may have built – and only looks at the harms related to people.

In the above, the term "hazardous situation" (Clause 3.35) refers to circumstances wherein people, property, or the environment are exposed to hazards. A "hazard", then, is a potential source of harm (Clause 3.9), and "harm", in general, is some physical injury and/or damage to the health of people, property, or the environment (Clause 3.8). Note that here the definition of harm is now wider than that used in the context of the software safety classification where only harm to people was discussed. Subsequently, software malfunctioning and causing a device to constantly overheat and thus shorten its lifetime might not be a factor in assigning the software safety classification, but if the decreased lifetime or device overheating caused harm to the patient it would feature in the risk management of the software. Similarly, for following green-coding practices and zero emissions goals the risk might be worth tackling even if no direct patient harm was involved. In practice, the incongruity here in the competing definitions of harm is appreciable but not earth-shattering.

Finally, the seriousness of harm is given a definition in Clause 3.22, where a serious injury is defined as an injury or illness that is a) life-threatening, b) results in permanent impairment of a body function or permanent damage to a body structure, or c) necessitates medical or surgical intervention to prevent such impairment or damage. Again, note that due to Clause 4.3a only harm to people should be considered while assigning a software safety class, but in practice, you will want to consider all types of harm in some adequate way in your overall risk-management operations.

Similarly, if interpreted strictly, the body of Clause 4.3a actually speaks of software contributing to an undesired outcome, not just a software failure contributing to it. The new flowchart added to the 2015 amendment in Figure 3, however, clearly speaks of software failure as the trigger. The wider focus appears to be reflected by regulatory expectations that your risk management also covers harm resulting from the correct use of the software within its intended use, as well as some foreseeable misuse of the software, in addition to harm caused by software failures.

4.3.3 Accept That All Software Will Fail

In discussing the assignment of the software safety classification, the IEC 62304 standard takes a rather pessimistic view that sooner or later all software will fail. That statement might conjure up images of the charmingly depressed robot Marvin from *The Hitchhiker's Guide to the Galaxy*, but here the point appears to be a solemn realization that you should not expect software to behave exactly as you programmed it. It may all be zeros and ones behind that code you wrote, but every once in a while, something unexpected will happen. It would be similarly appropriate to adopt the stance that eventually all hardware will fail too, but here the goal is to not get lulled into believing that software is perfect.

Accordingly, Clause 4.3a of the standard is adamant that in all cases the probability of a software failure occurring is set at 100%, in other words, all software will eventually fail. This does not, however, mean that all failures lead to harm. Many organizations have thus opted to split probability into two separate values, a P1 probability for a software failure (this is always 100%) and a P2 probability of that failure leading to harm (0–100%). Anyone familiar with probability mathematics will perhaps scoff at this solution as it does not change the outcome of the calculation, but the approach has proven practical in the real world. Note that while IEC 62304 discusses the matter in the limited context of software safety classification assignment, this approach is compatible with the instructions given in Annex C of ISO 14971:2019 concerning the actual risk-management activities (see Section 7).

4.3.4 Use of External Risk Controls

Clause 4.3a suggests that software with an initial safety classification of higher than class A may implement external risk controls to achieve a lower safety classification. Here, Clause 4.3a is adamant that only risk-control measures (mitigations) external to the software in question may be considered when assigning a safety classification to software (i.e., software system) or a software component (i.e., software item, software unit). These may involve redesigning the system architecture around the software component being classified via, for example, the following measures:

- Hardware measures
- Independent software system measures
- Healthcare procedures
- Other means to minimize the ability of the software to contribute to a hazardous situation

Note that independent software system measures are in fact considered external to the software in question. In other words, it is incorrect to assume that no risk-control measures could be implemented in software or that these would somehow not count in your software safety classification assignment or indeed the subsequent risk-management activities. Independence, however, is a key property here: if a failure in one place can lead to a failure elsewhere these two locations can't be considered independent. The rationale for segregation thus becomes equally important. Think of the bulkheads on the Titanic getting overflown by water, one by one as the ship tilted and sank, each flooded section weighing down the next and pulling it under. Think of that auxiliary power system that brought back the Terminator to finish that final fight after the glow in his eyes had already been extinguished. Neither system was totally independent of the component that failed in the first place, but the latter was independent enough to save the day. Another example here might be a software watchdog in a patient monitor that has been intended to ensure the software continues to run in a responsive manner. However, an overly aggressive timeout setting in the watchdog might cause the monitor to reboot unnecessarily, and thereby perhaps miss a critical event in the vital signs of the patient. In this case, the mitigation could thus cause the very harm it was intended to prevent.

Finally, a non-binding note to Figure 3 in Clause 4.3a offers commentary on risk-control measures implemented within a software by saying that the software may fail, this may then contribute to a hazardous situation, and the harm arising out of the failure may then also include the very harm the risk control measure was designed to prevent. This realization, too, harks back across a century to the above example provided by the Titanic.

4.3.5 Reducing Both Probability and Severity

A non-binding note to Figure 3 in Clause 4.3a clarifies that risk-control measures may reduce both the probability that the software failure results in harm and/or the severity of that harm. This too is a fundamental realization in addressing the risks involved in the use of your medical software, and one which is frequently asked about when discussing the risk management of medical devices.

Note that when using the P1xP2 approach introduced above it is not possible to lower the P1 probability as this is always set to 100% (or 1.0, depending on your notation) to reflect the assertion that software will always fail. It is thus left to us to consider reducing P2 via suitable means. Figure 4.2 below shows an example.

In the figure, the probability of occurrence of harm is controlled via mitigations such as including fault diagnostics. As a result of the risk control introduced the resulting risk is lowered to an acceptable level (not shown

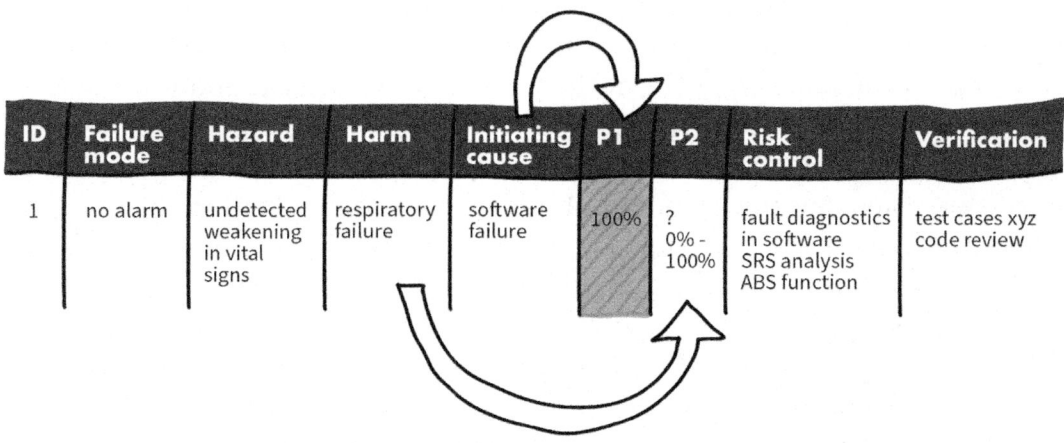

Figure 4.2 The use of P1 and P2 in a risk matrix.

in the figure). See Section 12.4 for more discussion of the Failure Mode and Effects Analysis (FMEA) methodology.

4.3.6 Recording and Inheriting the Classification

Clause 4.3d instructs that a child component (be that a software item of a software system, or a sublevel item underneath another item) will inherit the safety classification of the parent unless the manufacturer documents a rationale saying otherwise. The rationale must also explain segregation between the components so that they may be judged separate for classification. The resulting safety classification for each such component (i.e., software item) must be documented somewhere in your documentation if it differs from that of the parent component (Clause 4.3e). Remember that according to Clause 3.25 all levels of the software decomposition hierarchy – software system, software item, and software unit – may be considered a software item. Thus, the basic approach here should be to document the safety classification of each component if it is seen to be different from that of its parent. Remember that you must also provide a valid rationale for this decision, along with why the segregation between the items can be relied upon.

Similarly, the resulting safety classification for each software system must be recorded in the risk-management file (see Section 10). Clause 4.3g reiterates that this classification is always C until another software class is assigned. Note that the safety classification of software items other than the software system does not have to be in the risk management file per se (Clause 4.3e). In practice, you will want to record the classification and the logic for it somewhere, but this may, for example, be in the architectural design and not in a risk-management document. Beware of repeating yourself in several documents, though, as this will cause you gray hairs if the classification of some component then needs to be revised and this leads to editing an array of documents.

Finally, Clause 4.3f states that when evaluating the compliance of a group of software items to this standard, the highest-classified item in the group is taken as a yardstick – the implication being that the high-risk components can't hide behind the backs of the lower-risk items, not that low-risk items are suddenly treated as high-risk items. The requirements placed on the software (i.e., processes and tasks required) according to that highest safety classification will then be observed over the entire group. Alternatively, a rationale for using a lower classification may be documented in the risk management file.

4.3.7 Final Remarks

Overall, the ABC software safety classification is a great concept, but enthusiasm for it has been dampened by its equation by developers with not so much the level of concern, but the amount of documentation required for the software. The classification has jokingly been referred to as the ABC classification on the amount of documentation.

Some organizations have concluded that it's easier to document everything based on class C and thus avoid the need to keep track of any classification peculiarities or train developers to know two or three different pathways through the processes. This approach is perhaps further justified by the expansion of some requirements from higher classes down to class A software items in the 2015 amendment. Using a one-size-fits-all approach here may also make it more straightforward to adapt to any upcoming changes in safety classification due to new versions of the standard, but this may not always make sense or be the most optimal use of your resources. Flattening the classification also flies in the face of the general risk-based approach advocated by ISO 13485 in prioritizing your use of resources to make the maximum impact where it truly matters. As a mechanism to define requirements for different types of software the safety classification is, however, a sound approach and one which you can navigate based on your specific circumstances.

Also, remember that the standard does expect a software safety classification to be defined for both the entire software system and for the individual software items it consists of. This classification is to be appropriate based on the risks associated with the software, and it is generally expected that a clinician has been involved in the determination. If sufficient clinical expertise is already involved on the risk-management level, and this software safety classification is based on that activity, it may be debatable whether a clinical expert is separately needed here. In any case, the hook into your risk-management activities does mean that a mere statement of "this feels like a class B software" is not enough. Not having a classification, or not having an adequate rationale for it is a sure way to step on a mine during auditing.

In practice, you'll need to ensure that you have mapped your SOPs and product documentation to the expectations of the IEC 62304 standard, which it has laid out using the ABC classification. This book will help you accomplish this goal by discussing each area of the development process in turn. In terms of making the smallest incision into your product realization SOP, you could write that you only use class C from the classification but then you must be prepared to swim at the deep end of the standard. This is equivalent to going yachting on the USS Missouri, if not the USS Arizona – not to make light of a dreadful incident. This may not be optimal for using your resources strategically.

We expect Clause 4.3 to be clarified in later editions of the IEC 62304 standard. The scope of the application for the assertion of P1 to always be 100% is not entirely clear or helpful as it is occasionally perceived to also apply to Clause 7 and the overall risk management of the device even if this is not necessarily the intention of Clause 4.3 here.

4.4 Legacy Software

Clause 4.4 was added in the 2015 amendment to provide a standardized way for addressing legacy software despite differences in their development model and any trivial deficiencies in their documentation. Think of this as the Danny de Vito subclause to Clause 4, or the good cop in the duo, even if that slightly inverts the roles Schwarzenegger and de Vito play in their joint ventures. The aim of this subclause is, in any case, to bring you up to speed even if the circumstances surrounding your early life don't quite match up with those of your more fortunate sibling. You can overcome issues in your genes and upbringing if you are smart about it all, just like in that 1988 action comedy by Ivan Reitman.

In essence, the clause provides an alternative streamlined pathway for legacy software to achieve compliance with the IEC 62304 without meeting all the requirements in its Clauses 5 through 9 (stated in Clause 4.4.1). Instead, the manufacturer of such software must perform a gap analysis against certain clauses of the standard, and then act to close the identified gaps. The term legacy software is given an unequivocal definition in Clause 3.36 as medical-device software legally placed on the market, and still marketed today, but lacking sufficient objective evidence that it was developed in compliance with the current version of IEC 62304.

The discussion in the clause is divided into four topics: risk management-activities, gap analysis, gap-closure activities, and rationale for use of legacy software. These topics will be discussed next.

4.4.1 Risk-Management Activities

Building on from Clause 4.2, Clause 4.4.2 makes two further stipulations. Firstly, Clause 4.4.2a requires that the manufacturer assesses any feedback and post-production information, both from within and without their organization, regarding incidents and near-incidents. Secondly, Clause 4.4.2b requires that the manufacturer manages risks associated with the continued use of the software. The latter involves performing risk management activities that consider the following aspects:

- The integration of the software in the overall medical-device architecture
- The continuing validity of risk controls implemented as part of the software (incl. controls for each potential cause for the software contributing to a hazardous situation)
- The identification of hazardous situations (incl. the identification of potential causes for the software's contribution to these).

In other words, you must ensure that you are keeping up to date on how your software is performing by receiving feedback, assessing other post-market information, keeping abreast of any changes to the host device (if any), and ensuring your risk mitigation continues to be appropriate. The implication here is that you must meet the requirements both when bringing legacy software to the market, and over the long term of its use. The clause does not speak of the decommissioning of the software or its host device, but that too may be a subject to consider.

4.4.2 Gap Analysis

Clause 4.4 on legacy devices is not meant to give you a free ticket to comply with the standard without worrying about the other 79% of the standard. To analyze the gap between your existing status and the expectations of the standard you as the manufacturer are required to perform a gap analysis.

Gap analysis, in case you are not yet familiar with it, is a popular term for just about any situation where you need to look at where you are, where you need to be, and what steps you need to take to get there. It is probably one of the very first terms you'll come across when first getting into medical devices and as an answer to a question it is about as satisfying as admitting you have a problem. A gap analysis is a careful thought you must put in to assess the situation and figure out what needs to be done.

Clause 4.4.3 of the standard provides a helpful limitation on the gap analysis needed here by instructing you to look at how your available deliverables meet the requirements placed on them by the following clauses:

- Clause 5.2 on Software requirements analysis
- Clause 5.3 on Software architectural design
- Clause 5.7 on Software system testing
- Clause 7 on Software risk management

In the above, when the standard speaks of available deliverables don't be fooled into thinking that some deliverable that you don't have but is expected by the standard would not be a problem (although see Clause 4.4.3b below). You must meet the requirements of the standard, not the other way around. You are, of course, expected to utilize the ABC safety classification of your software when analyzing the requirements. The listed clauses themselves will be covered in detail in the following sections of this book (see Sections 5 and 7).

Together the above four clauses make up 21% of the standard, which will represent a sizable reduction in the list of requirements to fit into your brain's working memory. Do note, however, that the whole clause on risk management is included on the list even though it's featured as a separate requirement in Clause 4.4.2 above. Similarly, Clause 6 requirements on maintenance apply should any changes be required on the legacy software, and Clause 9 requirements on problem resolution may also be applicable sooner or later. The cherry on the cake is the requirement for a quality management system discussed above (Clause 4.4.1). This will mean that the walls of the carefully crafted legacy software silo may be breached at will if any cracks appear during audits, inspections, vigilance, or customer feedback.

For all the above reasons, we would not be content with reading the subclause on legacy software and thinking we are ready to push such products out onto the market. Every gap analysis should perhaps come with at least a stopgap consideration of fallback positions and countermeasures should you need to adjust your position later. Remember that IEC 62304 may only be a semi-voluntary test report, but the software you put out in the context of medical devices will also be a major point of discussion in any regulatory audits. Your business considerations may also change over time. It pays to think ahead, but this consideration will also go a long way to hardening your reasoning on any point, and that will have both internal and external benefits.

Clause 4.4.3 only places three immediate requirements on the manufacturer regarding the gap analysis. You are required to:

- Assess available deliverables for continuing validity (Clause 4.4.3a)
- Evaluate the potential risk reduction gained by generating any missing deliverables (incl. performing the associated activity) when gaps are found (Clause 4.4.3b)
- Based on the above evaluation, determine deliverables and activities to be run (Clause 4.4.3c)

The clause defines software system test records as the minimum level of deliverables (see Clause 5.7.5) it expects to see. As a further non-binding note, the standard also adds that the gap analysis should ensure any risk controls implemented in the legacy software are included in the software requirements.

4.4.3 Gap Closure Activities

You will probably have heard the viral Albert Einstein quote that if he had an hour to save the world he would spend 55 minutes of that defining the problem and then the remaining 5 minutes solving the problem. This is all well and good, but the quote says nothing about actually implementing the solution. It might thus be better to go out for one last pint than to leave the fate of the world in the hands of a theoretical scientist, as brilliant as he was. Clause 4.4.4 will have none of this impractical grandstanding as it insists that the gap analysis made above is also acted on. This is where after all the what-ifs and daydreaming of the previous clause we get to go to the mattresses, to do the work required. This is where after all that talk of melting the one ring in the fires of Mordor we start putting one foot in front of the other.

The clause requires the manufacturer to establish and execute a plan to act on the identified gaps. This plan may be combined with the software maintenance plan (see Clause 6.1 in Section 6), as the standard suggests in a non-binding note. The standard takes an admirably practical stance by stating that missing deliverables may also be created based on objective evidence, instead of running the associated activities retroactively (Clause 4.4.4a). What that objective evidence may be is not discussed by the standard, but, in practice, accumulated post-market surveillance data including information on similar products, maintenance history, and customer feedback may provide answers here.

The plan made must address problem resolution for any problems detected in the software or deliverables, and it must do this in accordance with Clause 9 (see Section 9). Finally, any changes to the software must be performed in accordance with Clause 6 (see Section 6). In other words, sailing on a legacy yacht is a relatively smooth business until it's not. As soon as you are faced with issues or want to make changes you are again a few steps closer to working with the full standard and all modern expectations. Rightly so, but this will cause some friction in maintaining legacy software out there on the seas.

4.4.4 Rationale for Use of Legacy Software

Finally, Clause 4.4.5 requires the manufacturer to document the version of the software and a rationale, based on the outputs of Clause 4.4, for its continued use. A non-binding note clarifies that fulfilling the requirements of Clause 4.4 enables further use of the software in accordance with IEC 62304.

Chapter 5

SOFTWARE DEVELOPMENT

Whatever your chosen software development model is – that classic waterfall so strikingly depicted in The Last of the Mohicans *or* Deliverance, *some flavor of the now popular agile model as perhaps neatly illustrated by the constantly mutating* Fast & Furious *saga, or something in between – this is where the in-depth software-specific requirements for it are located.*

By far the biggest clause of the IEC 62304 standard, Clause 5 (36% of the standard) dives deep into the actual bread and butter of the standard, the development of software. This is the heart of the standard. Here the standard sketches out the life cycle of software development from requirements to release – and even adds a preliminary step of development planning which does not feature in the definition of the term development life cycle earlier in the standard in Clause 3.24 but is very much a part of the software business.

The topics covered in this clause are given in Table 5.1 below. The table also notes which software safety classes each topic is applicable to and adds a final section on the aspects not covered by IEC 62304 but expected by ISO 13485. Note that the numbering of the following subsections corresponds to the top-level clause numbers in IEC 62304.

The clause immediately has apparent overlap with the ISO 13485 standard on quality-management systems, especially with the equally massive Clause 7 of that standard on product realization (34% of that standard). Clause 5 is also the first clause of the IEC 62304 standard where the software safety classification introduced in Section 4.3 is used to set different levels of requirements for low-risk and higher-risk software items. Note that

Table 5.1 Topics of Software Development

#	Topic	A	B	C
5.1	Development planning	⊠	⊠	☑
5.2	Requirements analysis	⊠	☑	☑
5.3	Architectural design	☐	⊠	☑
5.4	Detailed design	☐	⊠	☑
5.5	Unit implementation and verification	⊠	⊠	☑
5.6	Integration and integration testing	☐	☑	☑
5.7	Software system testing	☑	☑	☑
5.8	Release	⊠	☑	☑
5.9	The parts left out by IEC 62304	☑	☑	☑

☑ = all apply, ⊠ = some apply, ☐ = none apply

the safety classification may be applied to a software system, software item, or a software unit, and all of these are to be considered software items (note to Clause 4.3g).

Both standards talk of development stages in similar, compatible ways, but where the more generic ISO 13485 talks of design and development (D&D), thus remaining applicable to all sorts of products besides just software, the IEC 62304 standard speaks of software development specifically. IEC 62304 does occasionally refer to the wider system around the software and asks that software development occurs in sync with the development of the host device, but otherwise the details of developing non-software products are left outside the discussion.

Table 5.2 below shows a suggested mapping between the stages expected by either standard. Note that the purpose of the table is to identify related stages (and thus clauses) in the two standards, but not to suggest in any way that one standard would cover the requirements of the other.

The complementary nature of the two standards is particularly visible in the way IEC 62304 adds meat between D&D inputs (ISO 13485 Clause 7.3.3) and outputs (ISO 13485 Clause 7.3.4), and thus provides real structure to the development activities of software products. In other words, the ISO 13485 standard does not instruct how you proceed from D&D inputs to the outputs, it only wants to make sure the output makes sense considering the input, and that your reviews (including those based on your testing,

Table 5.2 Alignment of ISO 13485 and IEC 62304 Clause 5

Stage	ISO 13485	IEC 62304
Planning	7.1 Planning of product realization	5.1 Development planning
	7.3.2 D&D planning	
Inputs	7.3.3 D&D inputs	5.2 Requirements analysis
Design and implementation	*Surprise! The actual value-adding work that takes place between D&D inputs and outputs is not covered by the ISO 13485 standard.*	5.3 Architectural design 5.4 Detailed design 5.5 Unit implementation and verification 5.6 Integration and integration testing 5.7 Software system testing
Outputs	7.3.4 D&D outputs	*Elements of Clauses 5.1–5.8 apply here, and vice versa. This is explained in Sections 5.1–5.9 below.*
Review	7.3.5 D&D review	
Verification	7.3.6 D&D verification	
Validation	7.3.7 D&D validation	
Transfer	7.3.8 D&D transfer	5.8 Release

verification, and ultimately validation activities) confer this regardless of whether these are single big reviews or a series of smaller assessments.

The verification and testing activities in IEC 62304, on the other hand, are discussed iteratively from the level of software units all the way up to the level of software systems. These stages are related to the review, verification, and – perhaps to a certain extent – also the validation activities discussed by ISO 13485, but as the focus in ISO 13485 is on the final product these are best discussed separately. Remember that IEC 62304 does not actually cover the validation and final release of the final product (medical device), not even when the said product consists entirely of the software developed (Clause 1.2).

Here, between D&D inputs and outputs, is where you can pull in the experience from your past development work, maybe even from business areas other than medical devices, and build a subprocess that you know works for your organization and matches with the expectations of both standards. If you don't like the waterfall model of development, you can design your process here according to agile or whatever sound development model you have. Some flavor of agile is the future, that much appears to be certain in the development of software. ISO 13485 thus provides the bookends and IEC 62304 the bookmarks in between but no specific development model is forced on you in between by either standard, as depicted in Figure 5.1.

In the following, the discussion is structured around the clauses of the IEC 62304 standard. The requirements of each standard, and the relationship between these is discussed in detail. Each section then concludes with a synthesis of the requirements. Note that unless a specific software safety classification is stated in the following the requirements apply to all classes of software and software items.

5.1 Development Planning

Before you embark on a software development project some planning is involved. That is unless you are John J. Rambo, and you want to make up your way through that remote town as you go – if you do decide that is you, be prepared for some unnecessary fights, heated conversations, and even jail time along the way. Development planning is when you will get to outline your path from an idea to a finished product. Your horizons may never be as wide as during the planning stage, as Figure 5.2 illustrates, so enjoy this relative calm before the coming whirlwind.

As they stand, both standards acknowledge the need for planning. IEC 62304 sees planning as looking a little different for low-risk software than high-risk software, but mostly the requirements placed on development planning here are identical for all classes of software. ISO 13485, too, is compatible with the notion that something high-risk requires more planning and checking than something that is low-risk, but it does not provide a safety classification for your use here in figuring out what requirements apply and how. Instead, ISO

Figure 5.1 Both standards look to provide bookends and bookmarks for your work.

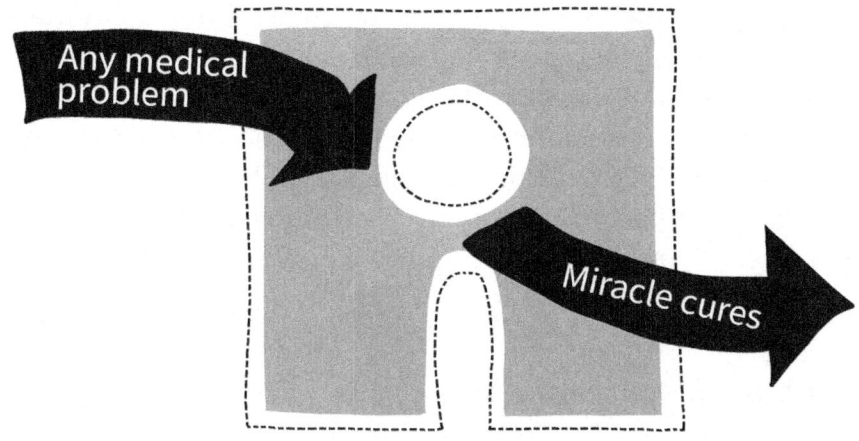

Figure 5.2 During planning your horizons are at their widest.

13485 talks of a risk-based approach that allows you to focus most of your resources on what matters the most (see Juuso 2022).

The relevant clauses of both standards are:

- ISO 13485 Clause. 7.1 Planning of product realization
- ISO 13485 Clause 7.3.2 Design and development planning
- IEC 62304 Clause 5.1 Software development planning

In the following, we will look at the expectations of both standards, one standard at a time, and then synthesize a solution to meet the requirements of both. As the backdrop for your software development is necessary in the QMS you have (Clause 4.1), we will start with ISO 13485 and then move to IEC 62304 in the discussion. The following discussion will identify which subclauses apply to which software safety classes.

5.1.1 Expectations from ISO 13485

5.1.1.1 Process Documentation

The baseline for software development planning is defined in Clause 7.1 of ISO 13485. Here you are required to have at least one process for product realization, and this is to also match consistently with the other processes of your QMS.

You must also incorporate risk management into the process, but the use of ISO 14971 is only suggested in a non-binding note to the clause here. Your regulatory environment may, however, have stronger feelings about the necessity of adopting the ISO 14971 standard in at least some of the operations. In general, the further away you get from the risk management for your product – e.g., from product risks to product-realization process risks to other process risks to general business risks – the lighter the weight of the standard should be. This is not to make light of risk management, it is certainly an essential part of everything you do regarding finished products, but it is not all that you need to know to set up your pipeline for medical-grade software development.

5.1.1.2 Planning Documentation

The name and format of your planning documentation is left up to you, but you are instructed to determine, as appropriate, the following:

- Product requirements (e.g., intended use, applicable regulations, quality objectives)
- Special resource needs (e.g., processes, documents, work environment, or personnel competence)
- Product-specific activities (e.g., verification and validation, monitoring, measurement, inspection, handling, storage, distribution, and traceability) and product-acceptance criteria

ISO 13485 adds additional requirements in Clause 7.3.2 addressing design and development planning. Here you are required to both plan and control the design and development of your product, and to ensure that planning documents are not only created but also maintained up to date throughout the process. Your plans must address your D&D stages (incl. review, verification, validation, and design-transfer activities), the overall D&D responsibilities (incl. authorities), and methods for traceability from D&D outputs to D&D inputs. This clause also calls for planning documentation to address relevant resource needs (e.g., personnel competence).

Clause 7.3.6 further clarifies that your verification plans must include methods, acceptance criteria, and as appropriate, statistical techniques with a rationale for sample size. Clause 7.3.7 takes the exact same stance on your validation plans. Remember that the standard here is also considering the manufacturing of physical products that are designed once and then churned out time and again off the production line. Unlike with software, with hardware products it will not be trivial to check whether each production run is identical to the last. With software products acceptance has more in common with the acceptance of the build or the original binary than with the acceptance of each clone known to be identical to its previous copies.

Note that both clauses, 7.3.6 and 7.3.7, stop short of requiring a documented process here and instead talk of working within your planned and documented arrangements. You are required to have procedures for D&D, but how much detail is in your SOPs may be debatable. In general, write your SOP on the level of your product portfolio, but plan to add necessary product- or project-level detail under that via planning documentation. You may also consider splitting your SOP into sections or even individual documents by the different types of products you make, e.g., software versus hardware. Your choice here will depend on your product portfolio, the number of each type of product-realization projects you regularly run, and the number of adjustments you expect to perform from one project to the next. In any case, keeping the SOP on an adequately high level should help you in your everyday operations. Keep in mind though that something too abstract will need to be translated into practice every time it is run, whereas something too micromanaging will be easier to follow but may lead to silently stifling the variation that occurs in real life. In other words, the former may cause sweating during the everyday and the latter during the last stages of development when hidden issues creep up. The sweet spot is somewhere in between.

Finally, in reviewing and approving the planning documentation you would do well to also ensure you meet the requirements of ISO 13485 for the conducting of D&D reviews as per Clause 7.3.5 (see Section 5.9.1). If you find that you need to revise some SOP that will be a process of its own (see Juuso 2022).

5.1.2 Expectations from IEC 62304

5.1.2.1 Process Documentation

Compared with the ISO 13485 standard, IEC 62304 is a little less concerned with a process definition and more concerned with a software development plan (or plans) to ensure appropriate development planning takes place in accordance with the requirements made in the whole of its Clause 5. You should thus have a process in place for your product-realization activities, but the IEC 62304 would be happy with this process being described in the planning documentation itself (e.g., in the software development plan as stated in the non-binding note 4 to Clause 5.1.1). IEC 62304 too appreciates the existence of a wider quality management system (QMS) around your software-development activities, and thus nothing prevents you from using processes defined in your QMS when developing your planning documentation. You are also required to map your software-development life-cycle model to the expectations of the standard, as explained in its introduction (see Section 3.2.2).

As a QMS is in fact required by the standard, and as that is likely to conform to ISO 13485, your best bet here is to ensure the standard operating procedures (SOPs) you have written based on ISO 13485 also answer to the requirements of IEC 62304 Clause 5.

5.1.2.2 Planning Documentation

IEC 62304 really wants to see a document called a *software development plan*. The name of the document can be something different, but you will be asked for such a document repeatedly so remember where you have the equivalent information in case you call the document something else.

This is basically your project plan for developing the software – although you may also have a separate higher-level software-system development plan if the two are not combined. The software development plan is principally instructed in Clause 5.1.1 and elaborated on all throughout Clause 5.1. The plan is given the following fundamental requirements.

- The plan must cover the activities of the clause in a way that is appropriate to the scope, magnitude, and safety classification of the software being developed.
- The plan must fully define or reference the software-development life-cycle model.

Furthermore, the plan must address the following aspects of the development project

- **Standards, methods, and tools** [c]
    The use of standards, methods, and tools associated with the development of class C software items (Clause 5.1.4, class C only).
- **Consideration of system requirements**
    The inclusion of system requirements as an input for software development (Clause 5.1.3a), and the procedures for coordinating system and

software development via, e.g., system integration, verification, and validation (Clause 5.1.3b). Note that the system requirements and software requirements may in some cases be the same (e.g., standalone software, software-only device).

- **Development processes employed**

  The development processes to be used (Clause 5.1.1a). Note that here a reference to your SOPs for product realization as per ISO 13485 should come in handy. Remember that you must also address the verification tasks required for each life-cycle activity (Clause 5.1.6b).

- **Deliverables expected and their verification**

  The deliverables (incl. documentation) of the activities and tasks (Clause 5.1.1b). Furthermore, the plan must identify those deliverables requiring verification, the milestones at which verification takes place, and the acceptance criteria needed (Clause 5.1.6a, c–d).

- **Documentation**

  The documentation produced during the development life cycle (Clause 5.1.8). This must include a) the title, name, or naming convention; b) purpose; and c) procedures and responsibilities for development, review, approval, and modification. Note that here a reference to your SOPs for document control as per ISO 13485 should come in handy.

- **Traceability**

  Traceability between requirements (both system-level and software-level), system tests, and risk-control measures implemented in software (Clause 5.1.1c). Note that IEC 62304 Clause 3.32 defines traceability quite narrowly as the degree to which a relationship can be established between two or more products of the development process. The definition given by ISO 9000, and thus used by ISO 13485, is compatible but also explicitly deeper: traceability refers to the ability to trace the history, application, or location of an object.

- **Risk management**

  The conducting of risk-management activities and tasks, including those related to SOUP components, in accordance with Clause 7 on risk management (Clause 5.1.7).

- **Configuration and change management**

  Configuration management and change management, also addressing SOUP components and software used to support development (Clause 5.1.1d). Also see the discussion of Clauses 5.1.9, 5.1.10, and 5.1.11 below, and Sections 6 and 8 of this book.

- **Problem resolution**

    Problem resolution covering the medical-device software, deliverables, and activities at each stage of the life cycle (Clause 5.1.1e). The plan must also address when the problem-resolution process is to be used (Clause 5.1.9f). Here too you might reference your SOPs as appropriate. Also see Section 9 of this book.

- **Integration** [B][C]

    A plan to integrate software items (incl. SOUP) and perform testing during the integration (Clause 5.1.5, classes B and C only). Note that it is acceptable to combine integration testing and system testing.

- **Avoidance of common software defects** [B][C]

    A procedure for the identification and avoidance of common software defects (Clause 5.1.12, classes B and C only). This must include identification of categories of defects introduced by the selected programming technology and relevant to the system (Clause 5.1.12a), and evidence of why these defects don't contribute to an unacceptable risk (Clause 5.1.12b). The standard offers IEC TR 80002-1:2009 as further reading on defects and hazardous situations. It may also make sense to invoke the use of good programming style guides, Integrated Development Environments (IDE), static code analysis tools, and perhaps even automated testing, as part of your answer to avoiding common defects. If you work on the same platform repeatedly, or develop similar software time and again, you may be able to reuse a previously made assessment here via a reference. You should still assess whether the present project poses any differences to those previous projects.

On the topic of traceability, note also that Clauses 5.2.6 and 7.3.3 discuss traceability. The former requires software requirements to be traceable to system requirements or another source (e.g., product specification, customer requirements), and the latter adds requirements for document traceability where the focus is on risk-management documents (see Section 7). The takeaway here is that traceability is an extremely important topic in medical devices and a relevant concern in just about all aspects of medical-software development. Making use of modern development tools will help in maintaining that traceability, but it is not enough to just think these will preserve links between everything as is needed. This is somewhat analogous to ISO 13485 document control, where using a good system (version management tool in that case) is expected, but you will still need to see to maintaining a practical version history over changes. Here we are concerned with tracking not just changes but inputs and outputs of various types. Good traceability is key to moving between such entities, going up and down, left and right, as needed to get the evidence you are looking for in making any informed decision, solving a nonconformity, or answering an audit question you happen to get. In the real world, traceability is often implemented well backward, but issues may occasionally pop up when trying to move forwards, or in the other direction less typically traveled. Think of this when planning your approach to traceability.

The standard adds a few non-binding notes on Clause 5.1.1 to clarify that your life-cycle model can specify different actions and deliverables for software items based on their safety class, that the actions may overlap, interact, and be performed iteratively or recursively, and that the actions may also be referenced from outside the plan (e.g., the SOPs of your QMS). All of these considerations will be useful to you in fashioning a lean, risk-based process for software development.

Clause 5.1.2 then echoes the requirements of ISO 13485 to keep your plan up to date as you progress. This does not, however, mean updating time stamps for the sake of getting new versions of a document. When your plans change you should update the planning documents, but to update the documents unnecessarily is just madness – and madness that causes more mayhem in your organization as people downstream try to figure out whether there have been changes and what effects these may have on their work.

Clause 5.1.9 goes into more detail on how the plan needs to address configuration management. Information here must include, or reference, the following.

- The classes, types, categories, or lists of software items to be controlled, and when these items are to be placed under configuration control (Clauses 5.1.9a,e).
- The activities and tasks needed in configuration management (Clauses 5.1.9b).
- The organizations responsible for performing activities, and their relationship with other organizations such as software development or maintenance (Clauses 5.1.9c–d).
- When the problem resolution process is to be used (Clause 5.1.9f)

Clause 5.1.11 requires that for class B and C software, all configuration items must be placed under configuration control before they are verified. Note that the requirement for this control to be documented was dropped in the 2015 amendment and that no requirement is thus made here to include a statement in the software development plan or link to a procedure.

Clause 5.1.10 requires that also those supporting items (e.g., tools, items, or settings) used in the development of class B and C medical-grade software must be controlled if these may impact the software. A further non-binding note offers compiler versions, make files, and batch files as examples of such items.

Note that most of the above requirements apply to all safety classes, with only the few stated exceptions being made. The topics required to be addressed may, in many cases, be addressed via discussion in the plan or by referencing another plan or your SOPs as needed. To simplify your work in defending the plan, and that of others in following it, it is generally a good idea to briefly set the context or otherwise explain why a reference is made and with what outcome instead of just providing a link. Beware of copy-paste and micromanagement, though.

5.1.3 Suggested Synthesis

The requirements of the two standards are complementary through and through. The quality-management standard is primarily concerned with having processes adequately described in the standard operating procedures (SOPs) of your QMS, and ensuring your D&D outputs match with what you wanted to get when you specified the D&D inputs. The software development standard relies more on a plan document to add a wealth of information it sees as necessary to ensure the success of the development project, although it too implies the existence of specific previously defined processes. Thus, defining your processes well enough on an appropriately high level in the SOPs will provide you with a good reference to cite in your plans, and save you from unnecessarily repeating or reinventing overlapping instruction in the plans. Similarly, the software development plan is your go-to plan for development, but it may then be augmented by other lower-level planning documents as is practical.

Table 5.3 provides a glance over the main requirements of both standards. Note the software safety classification used by IEC 62304 and remember that some clauses of ISO 13485 may be excluded with a valid rationale (see Juuso 2022). Also note that the table is an overview of any qualifying remarks presented in the previous discussion.

Table 5.3 Synthesis of Top-Level Requirements

Combined requirements	ISO 13485	IEC 62304 A	B	C
Process for development planning which	✓	✓	✓	✓
Matches with the other QMS processes	✓	-	-	-
Covers the entire development life cycle	✓	✓	✓	✓
Software development plan (SDP), incl./ref.	✓	✓	✓	✓
Product requirements	✓	-	-	-
Special resource needs	✓	-	-	-
Product-specific activities (incl. traceability)	✓	✓	✓	✓
Product acceptance criteria	✓	-	-	-
Reference to system design and development	-	✓	✓	✓
Standards, methods, and tools planning	-	-	-	✓
Integration and integration testing planning	-	-	✓	✓
Plan to integrate and test items (incl. SOUP)	-	-	✓	✓
Verification planning	✓	✓	✓	✓
Validation planning	✓	-	-	-
Risk management planning	✓	✓	✓	✓
Documentation planning	-	✓	✓	✓
Configuration management planning	-	✓	✓	✓
Control of supporting items	-	-	✓	✓
Configuration item control before verification	-	-	✓	✓
Avoidance of common software defects	-	-	✓	✓
Keep SDP updated	✓	✓	✓	✓

Based on the combined requirements, a synthesis between the standards may be negotiated based on the following items.

5.1.3.1 Standard Operating Procedure (SOP)

A documented process for development planning is required by ISO 13485 and vastly beneficial for consistently observing the requirements laid out by IEC 62304. This process can be neatly defined in the SOPs of your QMS – this is SOP-10 if you follow the QMS model set out in our previous book (Juuso 2022). In writing the SOP for product realization your objective here is to expand the earlier SOP to match the requirements of IEC 62304. Remember to refer to your other related processes (e.g., supplier management, infrastructure management, software validation) as is relevant, thereby avoiding the urge to repeat or reinvent overlapping instruction here.

This SOP will house your processes for software development in accordance with both standards. Here you should ensure the SOP instructs software-planning activities in accordance with the stages expected by both standards, including the key deliverables and reviews of each stage. In particular, the SOP should instruct the use and high-level structure of the software-development plan and allow the plan to add any further project-specific information. Don't plan to unnecessarily repeat information in the plan, but instead refer to the SOP as setting up the grand lines and add additional detail as needed – and as allowed by your description in the SOP.

The contents of the product realization SOP should correspond to Sections 5 through 9 of this book, specifically to what is discussed under each subsection on the suggested synthesis. To write the SOP, take your earlier version of SOP-10 (Juuso 2022), edit it to describe your software-development life-cycle model, add the deliverables suggested in the synthesis offered by this book, and add any further instruction you feel is warranted for your organization. Our suggestion is to structure the instructions in the SOP according to Sections 5 through 9 of this book and use each of these sections to aid in the writing of the corresponding section of the SOP. You may also split the instructions into several SOPs should this feel more practical to you, perhaps again along the lines of Sections 5 through 9. Remember to also address the software safety classification and legacy software from Section 4 as these are relevant to your operations.

5.1.3.2 System Development Plan

This is the highest-level development plan for an individual development project that is to take place under your QMS. The objective here is to describe the development of the whole system at a top level. The standard implies that you are to develop system requirements in or based on the plan, but this is poorly instructed in the standard. The system-level requirements are then referred to while developing the software under that system.

The standard also states, via a non-binding note to Clause 5.1.1, that this plan may be combined with the next-level development plan, the software-development plan (see next), particularly if the software is standalone (note to Clause 5.1.3b). This should be particularly appealing when the medical device to be developed is a software-only device, and thus the system requirements and software requirements may be the same. The standard suggests this combination of the plans to take place by including the software development plan in the system development plan, which makes great sense, but it then proceeds to place all the requirements on the software development plan in its later clauses. You may thus wish to call the combined plan by either name, or their combination, but you must be clear on the setup yourself if you combine these plans.

The whole existence of a system development plan is taken for granted by the IEC 62304 standard and not instructed well in any of its clauses. The note to Clause 5.1.3b and the overall taxonomy set up in Clause 3 (i.e., system, software system, software item, software unit) make it seem self-evident that you should indeed have such a document, and this should somehow mirror the content of the software development plan on a system-level. This is, however, not explicitly required by either standard. You will do well to consider the system-level beyond your software system (e.g., any hardware, environment, and users for your software), but whether you decide to do so in a separate plan or in the same plan you make for your software is a choice you can make. The system-development plan is not discussed to any great detail by either standard.

5.1.3.3 Software Development Plan (SDP)

Perhaps one of the most essential documents in your stack of documentation is the software development plan because it sets the foundations for all that is to come. The SOP already described your overall approach and process for software development so the objective here with the development plan is to add appropriate project-specific information, if that wasn't already added by a separate system development plan.

The software development plan is a key document for running the software-development activity. The plan is explicitly required by IEC 62304, and it acts as a logical extension of your product realization SOP as required by ISO 13485. As always, the purpose of a plan is to state what work you tackle and how – if the properties of this change in some meaningful way you should update the plan. No particular review of the plan is required by either standard, but make sure you follow your own documented process here to approve the plan and that you also follow ISO 13485 requirements for control of documents and records. If your plan is not approved at all, or if it is approved by a summer intern, that will raise eyebrows during any audit, but don't feel like every plan needs to be approved by the whole management team or the CEO either. Your management may greenlight a project and assign a project manager or a steering group, and then take a backseat until a later phase. The approval chain here will depend on you, but make sure you can demonstrate top management commitment on an appropriate level and that the chain of command is both clear and reflected in the checkups you instruct. The involvement of top management may, for example, be demonstrated through participation at some key reviews or through device marketing decisions and the setting of fundamental policies for topics such as risk management and cybersecurity.

The plan must be appropriate to the type of software being developed, and the safety classification of that software. To satisfy ISO 13485 requirements it should determine project responsibilities, resource needs, product requirements, and product-specific activities, as outlined above, unless those are covered by a separate planning document (e.g., project plan, system development plan).

The plan should address the entire development life cycle of the software, which implies meeting requirements sprouting up all throughout Clause 5. In doing so the plan should make use of the D&D stages expected by both standards (see Table 5.2), and impose an appropriate level of control over these, e.g., via stage reviews and the control of deliverables.

To satisfy IEC 62304 requirements the plan must address the topics listed above in Section 5.1.2. Some topics call for information to be recorded in or referenced by the plan, while on other topics you can answer the questions by referring to the processes defined in your SOPs – where the instructions are written on a high-enough level not to micromanage and yet to adequately spell out your overall key activities, tasks, and deliverables. ISO 14971 risk management is also implicated for use in product realization by both standards.

The final topic of joint interest in the context of development planning is traceability. ISO 13485 expects traceability to be maintained throughout all your operations, especially between the D&D outputs and D&D inputs. It also expects you to address any product-specific traceability considerations in the planning documentation. IEC 62304 asks to see traceability preserved between requirements, system tests, and risk-control measures. To meet both expectations your SOP and your development plan must be on the same page on traceability and together ensure that the necessary records can be navigated back and forth as needed.

Note that the topics of referencing system development (Clause 5.1.3), referencing standards, methods, and tools (Clause 5.1.4), integration and integration testing (Clause 5.1.5), configuration management (Clause 5.1.9), and avoiding common software defects (Clause 5.1.12) raised by IEC 62304, and discussed above, are not directly discussed by ISO 13485. Similarly, the topic of supporting items to control (Clause 5.1.10) is only discussed by ISO 13485 via infrastructure management and tool validation (see SOP-3 in Juuso 2022).

Don't unnecessarily repeat information here from the SOP, but instead refer to the process described in the SOP as setting up the grand lines and add any further project-specific information here. Where you draw the line between what is defined in the SOP and what is left for the plan is up to you. The SOP should provide the overall framework for your operations in a reusable way, but still leave enough room for adjustment on the planning level should the specifics of the software product require this. The plan should contain or reference all necessary information, and thus act as an access point into the development of your device.

It is not always possible to know everything in advance when writing the SOP, or even when writing the plan. To avoid a chicken-and-an-egg problem during planning, it is acceptable that not all of the information above is written in the plan with an equal sense of finality from the start of the development project. You may write in work items to investigate questions, and you may update the plan when developments call for adjustments to be made, but all the while you must make sure your plan is still up to date, makes sense, and is appropriate for the work going forward. In determining whether the plan is still up to date or not, we would look at whether the instructed course still holds true and not update the plan to record mile markers on that course. We have heard arguments to update the plan after a particular mile marker has been cleared, to record that event, but unless the passed mile marker changes the plan, we would not court the additional bureaucracy here. Update the plan when the plan needs to change, not as it is being followed. You may, of course, have an appendix or another referenced document indexing the status of the various documents your plan calls for. This way you can update that separate file without revising the whole plan, and you will have a ready-to-use dossier of your key files when you are compiling the final technical file.

5.1.3.4 Other Plans Mentioned

Clause 5.1 plays fast and loose with plans of different sorts, all of which are required to either be contained or referenced from the software development plan. It speaks of, e.g.,

- Tools planning [C] (Clause 5.1.4).
- Planning of integration and integration testing [B C] (Clause 5.1.5).
- Verification planning (Clause 5.1.6). Note here that ISO 13485 Clause 7.3.6 expects these plans to address methods, acceptance criteria, and (as appropriate) statistical techniques with rationale for sample size. See Section 5.9.2
- Validation planning (ISO 13486 Clause 7.3.7). Note here that ISO 13485 Clause 7.3.7 expects these plans to address methods, acceptance criteria, and (as appropriate) statistical techniques with rationale for sample size. See Section 5.9.3.
- Risk-management planning (Clause 5.1.7). See Section 7 and also SOP-8 in our previous book (Juuso 2022).
- Documentation planning (Clause 5.1.8). See Section 11 and also SOP-1 in our previous book (Juuso 2022).
- Configuration-management planning (Clause 5.1.9). See Section 8.

You might create individual plan documents for all these activities, but if your software development plan can already cover these on an adequate level why would you? You might also find that your SOPs already cover everything on a good level and there is not that much project-specific detail the plans need to add. You could thus more conveniently refer to the SOPs from your SDP and save your staff some head scratching in figuring out if the plan behind the copy-pasted text is the usual or something else. This consolidated, cascaded approach to planning will also hopefully save you from unnecessary work in maintaining multiple plans to all stay up to date. The one exception to the above is possibly your validation plan which holds the door for a critical and potentially costly activity that you do not want questioned after it has been completed. Here you may want to look at Section 12.2.3, and plan fastidiously before commencing with the validation.

If there is a need for the present software development project to deviate from the instruction in the SOP in any way, you may acknowledge those adjustments in the software development plan. Provided, of course, that the deviations are acceptable and allowed by the SOP. If your SOP says that your SDP (or some document addressed by it) must be chiseled in triplicate on stone tablets then that is what you must do; there is nothing the SDP can say to change that.

We would thus suggest adopting the SOP as the baseline for planning, then providing any known adjustments to that instruction for this particular project in that master plan, the software development plan. You might, for example, provide further information to consider during some development stage for this particular product, which you could then observe once reaching the corresponding step. If you then find that you need to develop a full-blown plan for some of the stages discussed, you should be free to do so within the window to set up your SOP and SDP. Crucially, you will not have tried to develop all these plans in-depth at the start when any small change in your planned trajectory could have thrown all the hard work on such other plans into the bin, and you won't have finalized an SDP with big gaping holes in it in terms of some of the above topics completely forgotten.

Just like a software item is understood by the standard, just about everything is always an onion you can peel to reveal new layers and provide more detailed instructions to guide the work as needed. Don't try to peel every onion in the software development plan, but know what onions you'll eventually need, and what you should perhaps consider when finally getting ready to peel those.

See Section 11 for an overview of what documentation and information is worked on during each stage of software development.

5.1.3.5 Use of Systems Instead of Documents

The IEC 62304 standard talks of the key types of information (e.g., requirements, architectural design, detailed design) to handle in a software development project. In many cases it is natural to think of all this information as leading to paper-like documents of different types (e.g., a specification, a design document) but you could also meet the requirements of the standard by maintaining this information in some database form inside a trusted practical tool. This way the information will be readily available to developers and a lot of the friction involved in maintaining such information over time will also be reduced.

Before opting for systems over documents, consider that you must still be able to prove what information was reviewed, approved, and thus cleared for use by activities (see Section 5.9.1). This may get particularly cumbersome if the underlying entries, such as individual requirements, exist in a free-flowing pool that may be added to at any time, but you are still required to ensure consistency across the whole set and assess the whole in some way for moving ahead to the next development stage. These issues may be solved using appropriately defined workflows in the systems, but also remember that you may then need to reproduce a particular snapshot of the information later, perhaps even years down the road. You may, for example, be asked to show the requirements or the architectural design used for your software 10 versions, or a decade ago, and prove that these formed a conformant set of information at the time. Retaining this information in the systems is not impossible, but over time as the systems themselves go through updates and bug fixes it may also be a more complex task than controlling documents might be. The sweet spot is perhaps in the use of a hybrid approach that lets some information evolve in systems but retains overarching assessments or key reviews in a more document-like manner.

5.2 Requirements Analysis

> *In the excellent Clint Eastwood film,* The Changeling, *the mother just wants her missing boy back. The police of 1930s Los Angeles think it's fine to respond by returning a boy almost matching the profile, albeit 3 inches shorter than expected. The mother is understandably distraught by the trade-in, as would any customer handed software that falls short of expectations. The requirements should not be treated as an alphabet soup, as illustrated in Figure 5.3.*

Developing requirements for the software to be built is a fundamental and profoundly important task. It is often claimed that most later issues detected with software can be traced back to faults in the requirements. This is to be expected as misunderstanding what it is you are to build, or losing track of this over time, intuitively takes effort to avoid. You might fail to grasp all the important aspects of the software at the start, or you might lose sight of these as requirements evolve over the development project. In both cases, luck would be your only life jacket to delivering a great piece of software once you spiral out of control. Paying enough attention to specifying requirements, on the other hand, may save you from floundering.

In practice, coming up with a solid list of input requirements here is both a science and an art. It may also be an iterative process where you approach

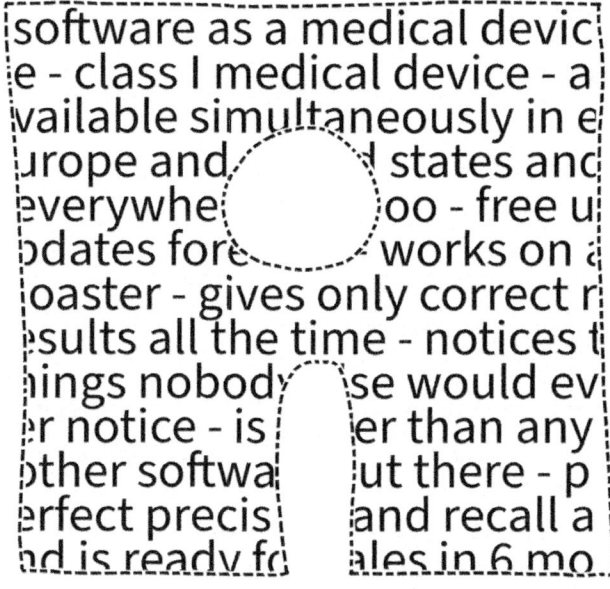

Figure 5.3 Requirements may be a bit of an octopus sushi to wrangle, but they should never be treated as an alphabet soup.

and refine requirements multiple times from different points of view as your understanding grows throughout the project. The two standards only provide inputs for you to carefully consider, and eliciting ready-to-use requirements from your customer is usually no small feat either. Much of your success in developing solid, complete requirements that do not require unnecessary changes later will come down to experience in your particular line of business. Whether you have a real or an imaginary customer – a talkative human being or a concept of a potential customer – to interview will also make a difference here.

The relevant clauses of both standards are:

- ISO 13485 Clause 7.3.3 Design and development inputs
- ISO 13485 Clause 7.2.1 Determination of requirements related to the product
- ISO 13485 Clause 7.2.2 Review of requirements related to the product
- IEC 62304 Clause. 5.2 Software requirements analysis

In the following, we will look at the expectations of both standards, one standard at a time, and then synthesize a plan to meet the requirements of both. The following discussion will identify which subclauses apply to which software safety classes.

5.2.1 Expectations from ISO 13485

The expectations for software requirements from ISO 13485 are primarily discussed in Clause 7.3.3 on design and development inputs, but also touched on in the customer-related Clauses 7.2.1 (Determination of requirements related to product) and 7.2.2 (Review of requirements related to product).

5.2.1.1 Inputs for Requirements

You must identify and maintain records of the inputs relating to the requirements of your product (Clause 7.3.3). The inputs must include the following:

- The intended purpose/use (incl. the functional, performance, usability, and safety requirements based on it)
- Applicable regulatory requirements
- Used standards
- Output of risk-management activities
- Experience from previous similar devices
- Any other requirements for the product or processes that you deem essential

Clause 7.2.1 further requires your organization (manufacturer) to determine requirements specified by the customer (incl. delivery and post-delivery activities), any user training needed to ensure specified performance and safe use, and any requirements not stated by the user but known to be necessary for the specified intended use.

5.2.1.2 Finalizing Requirements

You must review the inputs for adequacy and approve them after ensuring the requirements are complete, unambiguous, not in conflict with each other, and amenable to verification and validation. The standard also suggests the use of the IEC 62366-1 standard on usability as a further aid to consult here.

Clause 7.2.2 further requires that the product requirements are reviewed prior to the organization's commitment to supply the product to the customer (e.g., submission of tenders, acceptance of contracts/orders, or changes to these). The review of the product requirements is to ensure the following:

- Product requirements are defined and documented
- Any differences from previously expressed requirements are resolved
- Applicable regulatory requirements are met
- Any necessary user training is planned to be available
- The organization (manufacturer) can meet the defined requirements

You should also ensure that you meet the requirements of ISO 13485 for the conducting of D&D reviews as per Clause 7.3.5 (see Section 5.9.1).

5.2.2 Expectations from IEC 62304

IEC 62304 sees the final medical device as something that goes above and beyond the standard. The highest level it directly addresses is the software system of the medical device, i.e., the software component of the released medical device. The standard does at the same time concede in several clauses that occasionally the software system may in fact be the medical device (e.g., standalone software).

With the above mindset, Clause 5.2 requires the manufacturer to define and document software-system requirements – for each software system of the medical device – from the system-level requirements. The wider construct of the system, as defined in Clause 3.30 (see Section 3), is the high-level composite of the processes, hardware, software, facilities, and people relevant to meeting some objective or stated need. In other words, the system encompasses both the medical device and the context of its use in a very broad sense. The objective here is to define software-system requirements, but to do so you must also consider the wider system around it.

5.2.2.1 Inputs for Requirements

Clause 5.2.2 details the types of requirements the standard expects to see, provided these are relevant to the medical-device software. The standard acknowledges, in a non-binding note, that not all the requirements may be available at the start, and that the listed types of requirements (a–l) may overlap. Below we have grouped the requirements into seven practical collections for discussion.

- **Functionality and the environment**
    Description of what the software needs to do and under what circumstances. The clauses primarily addressed here are functional and

capability requirements (Clause 5.2.2a), and requirements related to IT-network aspects (Clause 5.2.2j). Requirements here may address, e.g., performance (purpose of the software, timing, etc.), physical characteristics and the intended computing environment (programming language, platform, operating system, hardware, network environment, etc.), compatibility needs (later upgrades, SOUP-components, other devices, etc.), and network issues (protocols, alarms, and handling of availability issues, etc.).

- **Data and interfaces to other systems**

    Description of what inputs, data, and interfaces the software needs to work with, what outputs it needs to produce, and how all this is stored. The clauses primarily addressed here are software system inputs and outputs (Clause 5.2.2b), interfaces with other systems (Clause 5.2.2c), and data definition and database requirements (Clause 5.2.2g). Requirements here may address, e.g., data formats and characteristics (value ranges, limits, defaults, etc.).

- **User interface and alarms**

    How the user should interact or use the software, and when should the software seek to interact with the user via alarms of different types. The clauses primarily addressed here are user interface requirements implemented by software (Clause 5.2.2f), and software-driven alarms, warnings, and operator messages (Clause 5.2.2d). Requirements here may address, e.g., the form of the user interface (full graphical, command line, API, etc.), and constraints on personnel. The standard suggests IEC 62366-1 (see Section 12.5) and IEC 60601-1-6 as further reading, both of which have been amended in 2020.

- **Security**

    What security concerns should be addressed and how in the software. The clause primarily addressed here is security requirements (Clause 5.2.2e). Requirements here may address, e.g., authorization, authentication, audit trails, system security/malware protection (incl. firewalls), communication integrity (incl. hash checks), and the protection of sensitive information. In the modern world, cybersecurity is definitely a part of the equation here, especially if your device interfaces with the outside world in some way.

- **Installation and delivery acceptance**

    How the software should be installed and how this should be verified. The clause primarily addressed here is installation and acceptance requirements of the delivered medical-device software at the operation and maintenance sites (Clause 5.2.2h). Requirements here may address, e.g., region settings for the keyboard, some database settings, time and date format, or relevant security requirements to ensure the safety, security, and essential performance of the functions performed by the software.

- **Operation and maintenance**

    How the software should be used and maintained, also by the user themself. The clause primarily addressed here is requirements related to methods of operation and maintenance (Clause 5.2.2i), and user maintenance requirements (Clause 5.2.2k). Requirements here may address, e.g., default presets for network protocols, procedures for adding new devices to the system, default values for cybersecurity settings, recovery in the event of disturbances or malfunction, and rollback requirements for updates or upgrades (see Section 12.2 on IEC 82304-1).

- **Regulatory requirements**

    What regulatory requirements are applicable here based on, e.g., the medical-device product being developed. The clause primarily addressed here is regulatory requirements (Clause 5.2.2l). Requirements here may address, e.g., the development of the software, the use of a quality-management system, the conducting of risk-management activities, and what approvals are needed for what types of software features.

Clause 5.2.3 (classes B and C only) states that those risk-control measures (mitigations) which are implemented in software are also included in the requirements as appropriate to the medical-device software. The standard acknowledges that these may not be known at the start of the development project, and that these may change over time. This may in practice be a cumbersome requirement to fulfill from the start as you may not yet know all your risks and their mitigations, and the safety classification of the affected software items may also still be veiled from you. Nonetheless, the requirement holds firm for B and C class software items.

5.2.2.2 Finalizing Requirements

Once software requirements are nearing completion, Clause 5.2.6 instructs how these are to be verified. No formal definition language is required here, but the requirements are to be verified and documented so that these:

- Implement system requirements (incl. those related to risk control)
- Can be uniquely identified

- Are expressed in terms avoiding ambiguity
- Do not contradict each other
- Are expressed in terms permitting the establishment and performance of tests (incl. acceptance criteria)
- Are traceable to system requirements or other sources

Clause 5.2.5 states that the manufacturer must also re-evaluate existing requirements (incl. system requirements) and update these as appropriate during the overall requirements analysis stage discussed here. Clause 5.2.4 further requires that the medical-device risk assessment is re-evaluated when software requirements are established, and it too is updated as appropriate.

5.2.3 Suggested Synthesis

The two standards are complementary. Table 5.4 provides a glance over the main requirements of both standards. Note the software safety classification used by IEC 62304 and remember that some clauses of ISO 13485 may be excluded with a valid rationale (see Juuso 2022).

As your software-development process needs to conform to IEC 62304 Clause 5.2, ensure here that your SOP for product realization discusses the development and review of requirements in accordance with the description above. Your software development plan may then add any necessary additional information.

Based on the combined requirements, a synthesis between the standards may be negotiated based on the following items.

5.2.3.1 Requirements Elicitation

IEC 62304 takes a fairly nuts-and-bolts view of requirements elicitation. It puts the pressure of arriving at an unambiguous and cohesive set of requirements squarely on the manufacturer, as indeed does ISO 13485.

IEC 62304 is concerned with largely technical requirements including data and interfaces, but it also acknowledges regulatory requirements face on, and implies a need to consider the user and the use environment when defining features, alarms, installation, and maintenance, for example ISO 13485, on the other hand, looks much more outside of the actual program code ISO 13485 places greater emphasis on listening to and understanding the customer.

Table 5.4 Synthesis of Top-Level Requirements

Combined requirements	ISO 13485	IEC 62304 A	B	C
Consider requirements from the intended use (purpose), including those not mentioned by the customer	✓	-	-	-
Consider use of standards	✓	-	-	-
Consider experience from previous similar devices	✓	-	-	-
Consider requirements for user training	✓	-	-	-
Consider other customer requirements	✓	-	-	-
Consider requirements from system requirements	✓	✓	✓	✓
Software requirements content	✓	✓	✓	✓
Include risk control measures in software requirements	✓	-	✓	✓
Re-evaluate medical device risk analysis	-	✓	✓	✓
Update requirements	✓	✓	✓	✓
Review/verify software requirements	✓	✓	✓	✓

5.2.3.2 Requirements Review (i.e., Design Input Review)

The expectations of the two standards on requirements development and review are highly compatible, but each has their own views on what aspects should be considered during the development and review of software requirements. You should review the requirements for completeness, unambiguity, traceability, and later verifiability via testing, but also in light of regulations, changed expectations, and your ability to meet the requirements.

In essence, this is the first of your three big reviews: your Design Input Review. The other two are covered in Section 5.8. Here you are making sure the requirements, i.e., the inputs to your development activities warrant moving ahead to the actual design phase.

Once software requirements have been established, IEC 62304 expects you to also re-evaluate and update higher-level system requirements and the medical-device risk assessment. ISO 13485 too expects the risk-management documentation to be up to date.

5.3 Architectural Design [B][C]

> *A classic in the disaster-movies genre,* The Towering Inferno, *opens with grandiose images of the world's tallest skyscraper designed by architect Paul Newman. In the film, Newman's character provides the grand strokes of the glass vision, but the engineers and economists have then been left to fill in the rest of the details to less-than-great results. The architectural plan sketching out that overall thing to build is also the goal fueling Clause 5.3 of IEC 62304 or sketching out the software symbolized by the floppy disk in Figure 5.4.*

Architectural design is where you translate those system and software requirements you have now established into a good bird's-eye view of the software you are developing. Here the ISO 13485 standard is out to get coffee and lets you and IEC 62304 iron out the details of how all this should be done. ISO 13485 still expects to later understand the key reviews you have performed here so make sure to document your reviews in a compatible way (see Section 5.9). Note that IEC 62304 does not require architectural design for class A software.

Clause 3.3 gives a formal definition for the term architecture as the organizational structure of a system or a component. This definition is quite broad and by itself would not give much reason to introduce a new term here. Architecture of software, or software architecture, is a common term, however, and one which instantly brings up images of software components organized neatly into grids like chocolates in a box – or like that dated 3D-animation at the end of Jurassic Park when the girl realizes that it's a Unix system.

The relevant clauses of both standards are:

- ISO 13485: None directly
- IEC 62304 Clause 5.3 Software architectural design

In the following, we will look at the expectations of both standards, one standard at a time, and then synthesize a plan to meet the requirements of both.

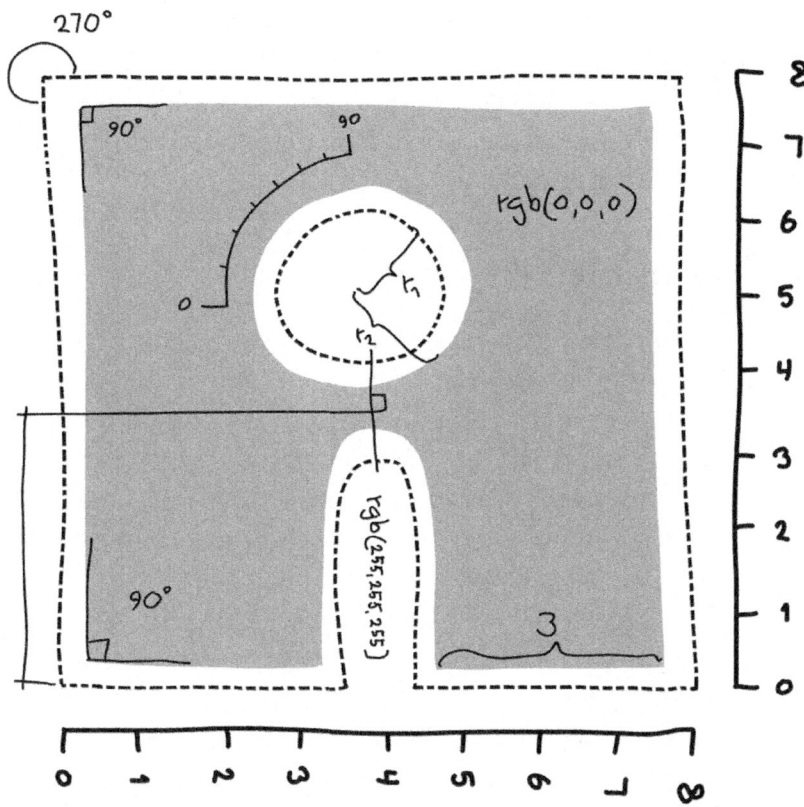

Figure 5.4 Floppies were once the modern equivalent to paper scrolls, but now they too may be relegated to ancient ruins on our architectural map.

5.3.1 Expectations from ISO 13485

Architectural design is a topic that does not appear in the ISO 13485 standard, but instead silently sits between its Clauses 7.3.3 (inputs) and 7.3.4 (outputs). The bookends the standard provides are between the D&D inputs and outputs so here you are in essence free to follow the IEC 62304 standard and translate that to your use.

5.3.3.1 Developing the Architecture Design

No requirements.

5.3.3.2 Verifying the Architectural Design

Despite ISO 13485 not placing any specific requirements on your architectural design or its review, you should still ensure that any review you make for the design meets the requirements of ISO 13485 for the conducting of D&D reviews in its clause 7.3.5 (see Sections 5.9.1 and 5.9.2). This much is expected by ISO 13485 if you later want to pull up or lean on your architectural design in any context recognized by the standard (e.g., conformity assessment of the final product).

5.3.2 Expectations from IEC 62304

The IEC 62304 expects you to design your software architecture according to Clause 5.3 if, and only if, your software safety classification is B or C. The standard does not expect architectural design to be completed for class A software. The idea here is that you can prioritize the software, and those software subsystems in your software that pose the highest risk and not jump through perhaps unnecessary hoops in the development of low-risk software items.

In practice, some organizations have adopted a one-size-fits-all model for software development, and thus chosen to design their software architecture for all software items using the same process they use for class C software. The upside here is that you won't have to train and maintain several different paths through development based on the safety classification, and you won't perhaps have to retroactively produce documents if the classification of some module changes. The downside is that you may be wasting some resources by using a howitzer to shoot peas if you are not careful – you may also be fooling yourself into thinking that a howitzer is a good instrument for shooting peas.

The expectations of architecture design are laid out in six short subclauses as discussed in the following. The top-level clause, Clause 5.1, has no content of its own. Note that aside from software item segregation, which is only required of class C software, all the other requirements apply to both B and C classes.

5.3.2.1 Developing the Architecture Design

The specific requirements for an architectural design are as follows.

- The design transforms the previously established medical-device software requirements into a documented software architecture that both describes the structure of the software and identifies the software items used in it (Clause 5.3.1). The architecture is also required to observe

any software system requirements (see Clause 5.3.6 below). Note that although the term software item technically may refer to software systems, software items, and software units – all of which are software items according to a non-binding note to Clause 3.25 – the requirement here does not extend down to the level of units. The requirement to identify software units is discussed along with detailed design in Clause 5.4.
- The design describes the interfaces between the software items, and also between them and any external hardware or software components (Clause 5.3.2).

In case a software item is identified as software of unknown provenance (SOUP; see Section 1.14), the manufacturer must also specify the following.

- The functional and performance requirements of the software item in relation to its intended use (Clause 5.3.3). Examples of functional requirements here might be the ability to parse XML messages over an interface, the string lengths allowed, the numerical space expected, and the supported character encoding when communicating over an interface. Similarly, examples of performance requirements might be performing some calculation in under 100ms or sorting data values at the same time.
- The system hardware and software requirements for the proper operation of the software item (Clause 5.3.4). Typical examples of requirements that could be made here are, e.g., processor speed, memory use, and input/output needs.

Some organizations have plunged in deep to define component-specific requirements to the above while others have been content with more generic performance-based requirements that are the same for all SOUP components. The choice is yours, but it will also depend on the criticality of the component and the risk associated with its function.

Finally, in case a software item is class C, the manufacturer is also required to identify effective segregation between software items as necessary for risk control (Clause 5.3.5). This also includes a statement on how the effectiveness of the specified segregation is ensured. Examples of how to ensure segregation may, for example, include running software items on separate, unshared resources. Note that this is a new subclause added in the 2015 amendment.

5.3.2.2 Verifying the Architectural Design

Finally, Clause 5.3.6 instructs the verification of the architectural design. Here the manufacturer must document verification that the architecture:

- Implements system and software requirements (incl. risk controls). Here the standard notes that a traceability analysis between the architecture and the underlying software requirements may be a practical tool but note that the unbinding advice does not cover system requirements
- Supports interfaces between software items, and between software items and hardware. Note that this subclause does not mention external software, but that was a part of the requirements in Clause 5.3.2 above
- Supports the proper operation of any SOUP items

The form of documenting this verification is up to the manufacturer. A simple review document identifying the usual details of exactly what was reviewed, by who, when, and to what effect is the basic expectation here.

The third requirement on the above list, item C, is interesting in that it may not be immediately clear what this entails. We are required to ensure the architecture we designed supports the use of the SOUPs as designed. If we think about our software as a battery-operated device and SOUPs as the batteries we put inside that device this intuitively translates to placing the batteries in correctly (with the right orientation of +/- poles) into the battery compartment. If the compartment door doesn't close, or if the batteries are in any order (plus pole to plus, minus pole to minus pole between batteries) there's a good chance the mental architecture inside our brain needs a little retuning. But what does this mean for actual SOUP items? The answer is that the architecture should make use of SOUP components as is appropriate for its goals and in agreement with how those components were intended to be used. Don't put an electric eel inside your boombox and expect good things to come.

As a sidenote, remember that your software development plan is expected by IEC 62304 to either contain or reference your verification planning. The most convenient way to address this on a universal level may be to reference your overall approach to the verification of process deliverables in your SOP. The verification work here for the architecture is a part of that overall verification activity for the software system.

5.3.3 Suggested Synthesis

As no specific requirements for architectural design are made by ISO 13485 the synthesis here is simple: follow the requirements made by IEC 62304. You should, however, ensure that you meet the requirements of ISO 13485 for the conducting of D&D reviews (see Section 5.9.1) and D&D verification (see Section 5.9.2).

Table 5.5 provides a glance over the main requirements of both standards. Note the software safety classification used by IEC 62304 and remember that some clauses of ISO 13485 may be excluded with a valid rationale (see Juuso 2022).

Did you notice how the requirements in this clause felt a little lighter than in the previous clauses? Here both standards seem to realize that you will have to do some real soul searching and heavy lifting to come up with the architecture of your new software, and thus give you some more room to breathe. The IEC 62304 standard, in particular, gives you some parameters for your work here, and the main properties you'll need to verify at the end, but otherwise you are given great leeway here to use the tools, models, and visualization concepts you wish. This is both a welcome and a very practical approach by the standard to, for once, conform to your elected operation models. In other words, you are free to do whatever you want as long as the outcome is correct.

Table 5.5 Synthesis of Top-Level Requirements

Combined requirements	ISO 13485	IEC 62304		
		A	B	C
Transform software requirements into an architecture	-	-	✓	✓
Develop an architecture for the interfaces of software items	-	-	✓	✓
Specify functional and performance requirements of soup item	-	-	✓	✓
Specify system hardware and software required by soup item	-	-	✓	✓
Identify segregation necessary for risk control	-	-	-	✓
Verify software architecture	-	-	✓	✓
Meet ISO 13485 requirements for verification review	✓	-	-	-

As your overall software-development process here needs to conform to IEC 62304 Clause 5.3, ensure that your SOP for product realization addresses the use of an architectural design on a high level. Your software development plan may then add any necessary additional information on developing and reviewing your design. If your design meets the requirements of IEC 62304 listed above, and your review of the design meets both the specific requirements of IEC 62304 and the general D&D review & verification requirements of ISO 13485, you are well-set for moving ahead.

5.4 Detailed Design [B] [C]

If Paul Newman got the role of the architect in the previous section, and the actual detailed design was left to monkeys behind the stage, the role of the person faced with the devil in those details was given to Steve McQueen as the fire chief battling that towering inferno. Similarly, here the aim is to dive inside that big picture and figure out what goes into it before it is all built. Figure 5.5 illustrates this.

Figure 5.5 Detailed design has us zooming into the architectural map we previously created.

This is the next depth for our diving bubble, or the next layer of the onion, after establishing an architectural design for your software. Here you will take your overall architecture and dive deeper into it to see all the way down to the level of software units (class B) and below (class C) – or skip this stage completely if you follow the class A route. In other words, note here that the IEC 62304 does not require a detailed design for class A software, and only requires a greatly simplified version of a detailed design for class B software.

The relevant clauses of both standards are:

- ISO 13485: None directly
- IEC 62304 Clause 5.4 Software detailed design

In the following, we will look at the expectations of both standards, one standard at a time, and then synthesize a plan to meet the requirements of both.

5.4.1 Expectations from ISO 13485

Detailed design is a topic that does not appear in the ISO 13485 standard, but instead silently sits between its Clauses 7.3.3 (inputs) and 7.3.4 (outputs). The bookends the standard provides are between the D&D inputs and the outputs so here you are in essence free to follow the IEC 62304 standard and translate that to your use.

5.4.1.1 Developing the Detailed Design

No requirements.

5.4.1.2 Verifying the Detailed Design

Despite ISO 13485 not placing any specific requirements on your detailed design or its review, you should still ensure that any review you make for the design meets the requirements of ISO 13485 for the conducting of D&D reviews in its clause 7.3.5 (see Sections 5.9.1 and 5.9.2). This much is expected by ISO 13485 if you later want to pull up or lean on your detailed design in any context recognized by the standard (e.g., conformity assessment of the final product).

5.4.2 Expectations from IEC 62304

5.4.2.1 Developing the Detailed Design [B,C]

IEC 62304 does not require a detailed design for class A software. For class B software all that is required is that the manufacturer refines the software architecture developed above by subdividing it into lower-level software items and software units until these are all represented as software units (Clause 5.4.1). Note that software units are by definition not divided further, and the standard also adds a non-binding note to say encountered software systems are not divided further here either. You should, however, address the encountered software systems via their own development documentation.

For class C software the following additional requirements are made.

- Each software unit will also receive a detailed design (Clause 5.4.2). This design is to be sufficiently detailed to allow the correct implementation of the unit.
- The detailed design for software units will document a design for any interfaces between software units and external hardware or software components. This design is to be sufficiently detailed to allow correct implementation of the unit and its interfaces.

5.4.2.2 Verifying the Detailed Design [C]

For class C software only, the detailed design is to be verified and documented to:

- Implement the software architecture (Clause 5.4.4a)
- Be free from contradiction with the software architecture (Clause 5.4.4b)

As with the architectural design above, here too the standard offers a helpful note to say that a traceability analysis between the detailed design and the underlying architecture may be a practical tool to meet the former requirement.

Note also that your software development plan is expected by IEC 62304 to either contain or reference your verification planning. The most convenient way to address this on a universal level may be to reference your overall approach to verification of process deliverables in your SOP. The verification work here is a part of that overall activity.

Table 5.6 Synthesis of Top-Level Requirements

		IEC 62304		
Combined requirements	ISO 13485	A	B	C
Refine software architecture into software units	-	-	✓	✓
Develop detailed design for each software unit	-	-	-	✓
Develop detailed design for interfaces	-	-	-	✓
Verify detailed design	-	-	-	✓
Meet ISO 13485 requirements for verification review	✓	-	-	-

5.4.3 Suggested Synthesis

As no specific requirements for detailed design are made by ISO 13485 the synthesis here is simple: follow the requirements made by IEC 62304. You should, however, ensure that you meet the requirements of ISO 13485 for the conducting of D&D reviews (see Section 5.9.1) and D&D verification (see Section 5.9.2).

Table 5.6 provides a glance over the main requirements of both standards. Note the software safety classification used by IEC 62304 and remember that some clauses of ISO 13485 may be excluded with a valid rationale (see Juuso 2022).

In other words, ensure that your SOP for product realization addresses the use of a detailed design on a high level and allows for more project-specific details to be added by the software development plan. In particular, pay attention to the software safety classification and how it affects it if and how the requirements apply to the detailed design of your software. Finally, review the detailed design if required.

After clearing the detailed design phase it's time to pat ourselves on the back and start implementing our designs. After one section of planning our approach to development and then three sections looking at the design of our software, we are now ready to roll up our sleeves and get to work in actually writing the software. This is where the script is written and it's time to call your cast out of their trailers to start making some movie magic.

5.5 Unit Implementation and Verification

The unit in The Dirty Dozen *may be put together from army rejects, but after Lee Marvin is finished with them, they are shown to be*

Figure 5.6 After all that designing, this is where we move to first doing and then checking the results.

> *capable of outshining the best-trained commandos in the theater of operations. Verification of this feat, too, is visible in the form of a blown-up chateau in northern France. The aim in Clause 5.5 is similar, only here the unit of study is the soldier being built to fit into the army. Figure 5.6 illustrates this.*

Software units are the atoms of your piece of software. From the point of view of the standard, these are software items that do not merit any further subdivision into software items. If you build these and integrate them together the way you wanted, you will end up with the entire software system you set out to develop. For class A all that is required is that you implement any units you intended to, but for class B and C, you must also verify that you got what you intended to get.

The relevant clauses of both standards are:

- ISO 13485: None directly
- IEC 62304 Clause 5.5 Software unit implementation and verification

In the following we will look at the expectations of both standards, one standard at a time, and then synthesize a plan to meet the requirements of both.

5.5.1 Expectations from ISO 13485

5.5.1.1 Unit Implementation

Unit implementation is a topic that does not appear in the ISO 13485 standard, but instead silently sits between its Clauses 7.3.3 (inputs) and 7.3.4

(outputs). The bookends the standard provides are between the D&D inputs and the outputs so here you are in essence free to follow the IEC 62304 standard and translate that to your use.

5.5.1.2 Unit Verification

Unit verification is a topic that does not appear in the ISO 13485 standard either. The topic of overall verification for your product, however, is something discussed in Clause 7.3.6 on design and development verification (see Section 5.9.2). Also remember what Clause 7.3.5 wants to see in terms of any D&D reviews made (see Section 5.9.1). You may choose to meet these requirements here head-on for software units but remember to then also address integration verification in the next stage and finally to ensure that you have covered your entire product through the body of your verification activities.

Alternatively, you might decide to treat software-unit verification as something separate from the D&D verification required by ISO 13485 and perform the latter separately. If you do so, realize that you may be forced to redo some verification activities if you then run into issues leaning on the earlier IEC 62304 verification from an ISO 13485 context. It therefore makes all the sense in the world to make your unit verification compatible with ISO 13485; the steps to do so are not steep. The few requirements set by ISO 13485 here are discussed in Section 5.9.2. An example of a potentially costly pitfall here is that ISO 13485 requires your verification to observe any interfaces or connections to other devices described in your intended use. In the case of software units, this will perhaps not matter much as you will be ensuring such connections are observed during your integration verification and software-system testing activities. It is nonetheless prudent to be aware of this assertion here.

In addition to meeting these verification requirements, you should also ensure that any unit-verification review you make meets the requirements of ISO 13485 for the conducting of D&D reviews in Clause 7.3.5 (see Section 5.9.1). This much is expected by ISO 13485.

5.5.2 Expectations from IEC 62304

5.5.2.1 Unit Implementation

Software-unit implementation is half the title of the clause, but it is only addressed by one short subclause: "the manufacturer is required to implement each software unit" (Clause 5.5.1). In other words, all the software units you intended to have must be implemented. In fact, the requirement applies to all classes of software and thus is not limited by the software units specified in the detailed design (class C software) or the corresponding refined software architecture design (class B software). For class A software, however, it may be difficult to check that you have indeed implemented all the units you planned to have as neither of these records is required to be available, but that is nonetheless the stated requirement here. In practice, you probably wouldn't start implementing even the flimsiest of class A software systems without some sort of a plan, software-item roster, or a stack of napkins. You should thus have some record of the software items you planned to have, and here the objective is to ensure you have not left any of those software items behind. Along the way you may have decided that Private Ryan was no longer needed for your campaign, and thus released him to go home, but if, on the other hand, you had just misplaced him or forgotten about him you should now take note and act accordingly. The purpose of this check is to ensure, in part, that you have done all that you intended, which is ultimately going to be a part of ensuring all of your activities are complete as ISO 13485 also insists.

To play devil's advocate here, note also that the term used here by the standard is software unit and not software item. In other words, you are required to implement all the units you planned, but you don't have to implement all the higher-level items. A unit is a type of an item, but an item is not necessarily a unit, after all. You shouldn't flirt with Dante's inferno though: always keep your documentation as up-to-date as you can.

5.5.2.2 Unit Verification [B C]

For classes B and C, the requirement here is to define a verification process (Clause 5.5.2) and acceptance criteria (Clause 5.5.3), and then perform documented verification (Clause 5.5.4). Note that class A software is exempt from all the verification requirements discussed in this section.

The process does not have to be described in an SOP – it could, for example, be instructed in your software development plan – but the manufacturer is required to establish strategies, methods, and procedures for verifying software units. If you perform verification by testing, you must also evaluate the adequacy of the test procedures. This evaluation may be part of the review and approval of your SOP (if appropriate), it may be recorded at the same time as approving your software development plan, or it may be a separate review.

The acceptance criteria for software units must also be defined, and the meeting of those criteria ensured, prior to integrating the units into any larger items as appropriate (Clause 5.5.3). Note that the standard here stops short of saying that the acceptance criteria should be set before the units are ready and their acceptability can be assessed, although it is generally a good approach to define what you want to get before you see what you got. The acceptance criteria could be based on, e.g., whether risk controls are implemented, if the unit matches the interface design set for it, and if the unit meets the programming guidelines and conventions of your organization.

Clause 5.5.4 (class C only) further requires that if the following aspects are relevant to the design, the manufacturer must include additional acceptance criteria to cover them:

- Proper event sequence (Clause 5.5.4a)
- Data and control flow (Clause 5.5.4b)
- Planned resource allocation (Clause 5.5.4c)
- Fault handling (error definition, isolation, and recovery) (Clause 5.5.4d)
- Initialization of variables (Clause 5.5.4e)
- Self-diagnostics (Clause 5.5.4f)
- Memory management and memory overflows (Clause 5.5.4g)
- Boundary conditions. (Clause 5.5.4h)

Finally, Clause 5.5.5 requires that the manufacturer must perform software-unit verification and document the results. Again, this requirement does not apply to class A. Note also that your software development plan is expected by IEC 62304 to either contain or reference your verification planning. The most convenient way to address this on a universal level may be to reference your overall approach to the verification of process deliverables in your SOP. The verification work here is a part of that overall activity.

5.5.3 Suggested Synthesis

As no specific requirements for unit implementation and verification are made by ISO 13485 the synthesis here is simple: follow the requirements made by IEC 62304.

Table 5.7 provides a glance over the main requirements of both standards. Note the software safety classification used by IEC 62304 and remember

Table 5.7 Synthesis of Top-Level Requirements

Combined requirements	ISO 13485	IEC 62304		
		A	B	C
Implement each software unit	-	-	✓	✓
Establish software unit verification process	-	-	✓	✓
Software unit acceptance criteria	-	-	✓	✓
Additional software unit acceptance criteria	-	-	-	✓
Software unit verification	-	-	✓	✓
Meet ISO 13485 requirements for verification review	✓	-	-	-

that some clauses of ISO 13485 may be excluded with a valid rationale (see Juuso 2022).

The process for your unit verification may conveniently be defined in your product realization SOP, and then any project-specific details added in the software development plan. You should address your overall approach to D&D verification in your process, i.e., explain how your unit verification activities fit into the verification of the entire product. Also ensure that your verification process and unit-acceptance criteria meet the requirements of IEC 62304 discussed above, and that your process includes some checks to ensure you have implemented all software units appropriate to your software safety class. As before, you should also ensure that you meet the requirements of ISO 13485 for the conducting of any D&D verification and reviews (see Sections 5.9.1 and 5.9.2).

5.6 Integration and Integration Testing [B][C]

As a film it's perhaps a guilty pleasure, but the Three Amigos! *starring Steve Martin et Co. is a serviceable example of uniting the disparate villagers for a common purpose. Ridding the village of a scourge may also be appropriate for a discussion of medical-software development. A more street-credible reference here may be Yul Brunner et Co. in* The Magnificent Seven, *or Akira Kurosawa's* Seven Samurai, *but the end result is the same. Here instead of integrating oppressed plebs and pitchforks, you are assembling the whole of your software. Figure 5.7 illustrates this.*

Figure 5.7 This is that "E pluribus unum" moment: out of many becomes one.

You may think of integration as stacking cargo containers when loading them on to a ship but remember that any interconnecting locks or other interfaces of those containers also need to be hooked up. What happens inside the containers was already implemented and verified above in Section 5.5, but now you need to build up from those blocks to the level of the whole you set out to develop and ensure that the pieces click together as you planned. The verification of this may be a simple inspection, as the standard suggests in a note, and the final testing of the whole is yet to come. Here the focus is on correct assembly as per your plans; later you'll ensure that the whole holds water. Note that it is, in fact, acceptable to also combine integration testing and system testing (see Section 5.7).

The relevant clauses of both standards are:

- ISO 13485: None directly
- IEC 62304 Clause 5.6 Software integration and integration testing

In the following, we will look at the expectations of both standards, one standard at a time, and then synthesize a plan to meet the requirements of both.

5.6.1 Expectations from ISO 13485

Software integration, verification of that integration, and then performing integration testing are topics that do not appear in the ISO 13485 standard, but instead silently sit between its Clauses 7.3.3 (D&D inputs) and 7.3.4 (D&D outputs). The bookends the standard provides are between the D&D inputs and outputs so here you are in essence free to follow the IEC 62304 standard and translate that to your use.

You may choose to meet your overall verification needs here head-on for the integration but remember that you may already have addressed unit verification during the previous stage, and that you will still need to ultimately ensure that you have covered your entire product through the body of your verification activities. Alternatively, you might decide to treat software verification as something separate from the D&D verification required by ISO 13485, and perform the latter separately, but realize that you may be forced to redo some verification activities if you then run into issues leaning on the earlier IEC 62304 verification from an ISO 13485 context. It therefore makes all the sense in the world to make sure your integration verification is compatible with ISO 13485. The steps to do so are not steep. The few requirements set by ISO 13485 here are discussed in Section 5.9.2. An example of a potentially costly pitfall here is that ISO 13485 requires your verification to observe any interfaces or connections to other devices described in your intended use.

In addition to meeting these verification requirements, you should also ensure that any integration verification review you make meets the requirements of ISO 13485 for the conducting of D&D reviews in its clause 7.3.5 (see Section 5.9.1). This much is expected by ISO 13485.

5.6.2 Expectations from IEC 62304

5.6.2.1 Integration Plan [B C]

Clause 5.1.5 (classes B and C only) calls for an integration plan to be included or referenced from the software development plan, but nowhere does the standard actually make requirements on the contents of this plan. The minimum answer might therefore consist of what is instructed by your SOP and what special considerations your software development plan may bring up for this particular project. In terms of describing your planned integration here, you should refer to the architectural design and detailed design, provided that these exist. We would argue that this set of documentation already addresses your planned whole and your arrangements for integration sufficiently.

If, however, you find that a separate in-depth integration plan is needed, you might develop that here before proceeding to the actual integration work and integration testing. If you find yourself writing an integration plan at the same time as the integration and verification work, you should reconsider your process.

5.6.2.2 Integrating Software Units [B C]

Clause 5.6.1 requires the manufacturer to integrate software units in accordance with the integration plan made in, or referenced from, the software development plan. In other words, the standard wants to see the reality of what you develop match what you planned. Should this not be the case you may need to go back a few squares and revise your plans.

5.6.2.3 Integration Verification [B C]

To ensure that all the blocks have been put into place as planned, the manufacturer is instructed to perform verification of the software integration (Clause 5.6.2). Clause 5.6.2 also makes it clear that integration here addresses all the blocks you described in your integration planning, whether the software units build software items or the software system itself.

Clause 5.6.2a doubles down on the above message and requires the manufacturer to verify that the software units have been integrated into software items and the software system. Clause 5.6.2b applies even more superglue around this requirement by stating that also the integration of hardware items, software items, and support for manual operations of the overall system is verified. By support for manual operations the standard refers to, e.g., human-computer interface features such as voice control and online help menus.

In a non-binding note to the clause, the standard offers an opinion that the verification activity here most likely takes the form of some of inspection, and that the purpose of the verification is limited to ensuring the integration has been done as planned. In other words, this clause is not concerned with what happens inside the cargo containers, or even if the connected set of cargo containers works as intended, only that the containers have been stacked and connected as planned. The test of the whole is chiefly addressed in the next clause (see Section 5.7).

Whatever form the verification takes, there must be records of it, as always. Note also that your software development plan is expected by IEC 62304 to either contain or reference your verification planning. The most convenient way to address this on a universal level may be to reference your overall approach to the verification of process deliverables in your SOP. The verification work here is a part of that overall activity.

5.6.2.4 Integration Testing [B][C]

If the verification of the integration instructed above was just a crude inspection that the integration plan was implemented – that the building instructions for your Lego set were followed – here you will get to make sure the thing you built works as intended. If you built a paperweight according to the instructions of a Lego digger this is where you should notice your mistake. If the vast cargo ship you stacked full of containers gets stuck in the Suez Canal that probably is not your fault, but if some of the crates fall off because you neglected to observe the loading plan you will feel the wrath of those looking on.

Clause 5.6.3 requires the manufacturer to test the integrated software items in accordance with the integration plan and record the results. The objective is to address whether the integrated software items perform as intended (Clause 5.6.4). The standard provides a helpful list of aspects to consider here in a non-binding and non-exhaustive note to the clause:

- The required functionality of the software is achieved
- Risk controls are implemented
- Specified timing and other behavior is reached
- Internal and external interfaces function as specified
- Testing under abnormal conditions (incl. foreseeable misuse)

A second, deceptively diminutive note to the clause then points out that integration testing may in fact be combined with overall software-system testing (see Section 5.7) into a single set of planned activities. This profound statement should give you welcome flexibility in matching the expectations of the standard with your development pipeline. Wherever you see your integration testing take place, the odds are that you will apply some form of continuous integration (CI) methodology and use automated tests conducted as part of merge/pull requests.

Clause 5.6.5 requires that the manufacturer evaluates the integration test procedures for adequacy. Appropriate regression tests are to be used to demonstrate that defects have not been introduced into previously integrated software when integrating software items (Clause 5.6.6). Note that regression testing itself is defined in Clause 3.15 as the testing required to determine that a change to a system component has not adversely affected functionality, reliability, or performance, and has not caused new defects.

The records kept of the integration testing must include the following (Clause 5.6.7):

- The test result (pass/fail and a list of any anomalies discovered)
- Sufficient details to enable the test to be repeated later as needed, e.g., test-case specifications (incl. required actions and expected results) and identification of the test environment (incl. equipment and software used)
- The identity of the tester

Note that the term anomaly is given a formal definition in Clause 3.2. Here the definition is quite loose referring to any condition that deviates from the expectations set up by requirements specifications, design documents, standards, or a person's perceptions and expectations. In addition to testing, anomalies are said to arise out of review, analysis, compilation, and the use of the product or its documentation.

Finally, Clause 5.6.8 requires that any anomalies discovered during integration and integration testing are fed into a software-problem resolution process (see Section 9).

5.6.3 Suggested Synthesis

As no specific requirements for integration, integration verification, and integration testing are made by ISO 13485 the synthesis here is simple: follow the requirements made by IEC 62304. You should, however, ensure that you meet the requirements of ISO 13485 for the conducting of D&D reviews (see Section 5.9.1), and that your verification activities here are compatible with the requirements for conducting D&D verification (see Section 5.9.2).

Table 5.8 provides a glance over the main requirements of both standards. Note the software safety classification used by IEC 62304 and remember that some clauses of ISO 13485 may be excluded with a valid rationale (see Juuso 2022).

In other words, ensure your SOP for product realization addresses unit integration, integration verification, and integration testing (incl. evaluation of the test procedure) on a high level, and allows for more project-specific details to be added by the software development plan. Your process should then ensure that you have integrated all the software units, verified this

Table 5.8 Synthesis of Top-Level Requirements

Combined requirements	ISO 13485	IEC 62304		
		A	B	C
Integrate software units	-	-	✓	✓
Verify software integration	-	-	✓	✓
Software integration testing & testing content	-	-	✓	✓
Evaluate software integration test procedures	-	-	✓	✓
Conduct regression tests	-	-	✓	✓
Integration test record contents	-	-	✓	✓
Use software problem resolution process	-	-	✓	✓
Meet ISO 13485 requirements for verification review	✓	-	-	-

according to the integration plan (included in or referenced from the software development plan), performed integration testing, and evaluated the test procedures for adequacy, as required.

5.7 Software-System Testing

This is where the marriage between the two standards gets interesting. This is where Mrs Smith takes a sniper rifle to Mr Smith in the desert, where Mr Smith rushes home to take stock of the pair's armory, and the two then battle out their world views as Brad Pitt and Angelina Jolie did, first on the silver screen in that film and then apparently more amicably at home. Figure 5.8 illustrates this.

The two standards look at the same thing – software in this case – from two different angles and want to ensure it all makes sense according to their points of view. IEC 62304 logically looks at the forest and the trees that make up the forest. ISO 13485 equally logically looks at whether your forest meets the requirements you had for a forest and the expectations your customer now has for a forest.

This inspection is to take place via testing on many levels as IEC 62304 sees it, but also for the purposes of verification (what you wanted) and validation (what your customer needed) as ISO 13485 expects. See Section 1.14 for more discussion on the difference in the point of view between the

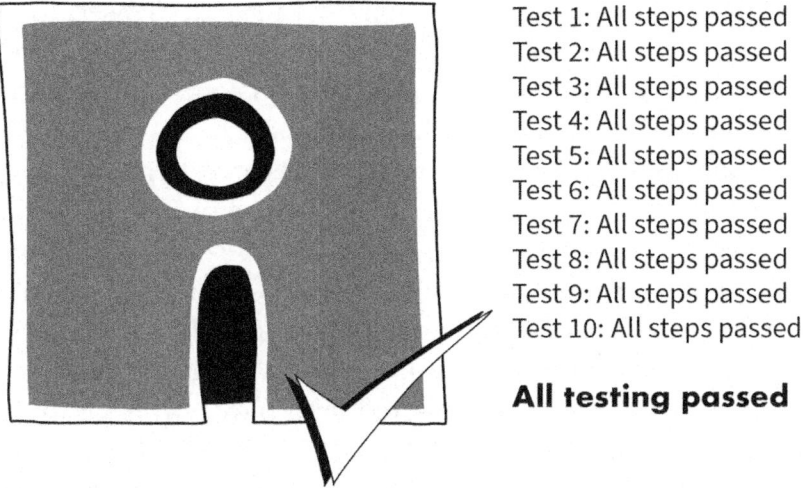

Figure 5.8 Testing should lead to that "all systems are go" feeling.

standards. Also note that, according to IEC 62304, software system testing was previously only required of higher software safety classes, but since the 2015 amendment the requirements here apply equally to all classes of software.

The relevant clauses of both standards are:

- ISO 13485: None directly
- IEC 62304 Clause 5.7 Software system testing

In the following we will look at the expectations of both standards, one standard at a time, and then synthesize a plan to meet the requirements of both.

5.7.1 Expectations from ISO 13485

The requirements ISO 13485 places on the testing of your software are addressed via the outputs, verification, and validation requirements presented in its Clauses 7.3.4 on D&D outputs, 7.3.6 on D&D verification, and 7.3.7 on D&D validation. The lens taken by the standard is markedly different from the bifocal lens of a decomposed lower level (software units and items) and an integrated top level (software system and system) taken by IEC 62304. If you follow the structure of software development instructed by IEC 62304 you will also have seen to many verification activities already. It would be a shame to have to repeat work just because ISO 13485 has a somewhat different line of sight. You may not have addressed the validation activities to an equal level of care, but you may already have a good foundation for performing that work if you have a customer in the loop and a thought-out intended use for your planned product.

To make sure no work is unnecessarily replicated here, our suggestion is to perform software-system testing as instructed by IEC 62304 and then, once the software system is ready, perform any necessary additional checks on the level of the final medical device to tie up your verification activities and perform the important validation activities. In other words, look at what you have accomplished on the parts-level, what you'll still need to test on the device-level in order to feel confident, and move on from there.

Aside from verification and validation activities, which we will sidestep for the moment, the ISO 13485 standard may be left to the sidelines in terms of software-system testing. As before, you should nonetheless ensure that any review you make here meets the requirements of ISO 13485 for the conducting of D&D reviews (see Section 5.9.1) and that you observe the ISO 13485 requirements for conducting D&D verification (see Section 5.9.2) if the verification here is to feature in the verification of your overall product.

After finishing with testing here we will move to software release in the next section. The discussion there will remind you of the ISO 13485 requirements related to review, verification, and validation before release.

5.7.2 Expectations from IEC 62304

As noted above, the requirements for software-system testing apply equally to all software safety classes since the 2015 amendment of IEC 62304. The requirements are given in Clause 5.7 of the standard.

5.7.2.1 Establishing and Performing Testing

The manufacturer is required to establish and perform testing against all software requirements (Clause 5.7.1a) and to evaluate the adequacy of the verification strategies and test procedures (Clause 5.7.1b). Clause 5.7.1 also provides a minimum set of information to establish for tests:

- Input stimuli
- Expected outcomes
- Pass/fail criteria and procedures

A non-binding note to the clause suggests that in addition to separate tests for each requirement, tests covering combinations of requirements may be performed. Another particularly helpful note adds that software-system testing may be combined with integration testing, and that it is even acceptable to test software requirements in earlier phases. Adopting the latter advice may speed your overall process greatly but remember to somehow still take ownership of the whole and ensure that the whole system has been tested appropriately.

Clause 5.7.2 instructs that any anomalies discovered during software-system testing are fed into a software problem resolution process (see Section 9).

Finally, Clause 5.7.3 requires that when changes are made during testing the following actions are taken:

- The effectiveness of the change in correcting the problem is verified by repeating tests, and performing additional or modified tests (Clause 5.7.3a)
- Conduct testing to demonstrate that unintended side effects have not been introduced (Clause 5.7.3b)
- Relevant risk-management activities are performed as defined in Clause 7.4 (Clause 5.7.3c). See Section 7.4

The minimum contents of software-system testing records are defined by Clause 5.7.5 so that the repeatability of the tests is ensured. The information to be covered includes:

- The version of software tested, the date of testing, and the identity of the tester (i.e., the person responsible for executing the test and recording the results)
- Identification of the test environment (incl. relevant hardware and software test configurations, and relevant test tools)
- A reference to test case specifications (incl. test case procedures, required actions, and expected results)
- The test result (pass/fail and a list of any anomalies discovered)

Note that the terms used here differ somewhat from those instructed above in the clause for integration testing, but do not do so in any material way in terms of content. The terms used by this clause are given in the brackets above, but the more general terms are otherwise used for consistency with the above clauses.

156 ■ *Medical-Grade Software Development*

5.7.2.2 Verification of System Testing

Clause 5.7.4 requires the manufacturer to evaluate the appropriateness of system-testing verification strategies and test procedures. The objective of the verification here is to verify that:

- All software requirements have been tested or otherwise verified (Clause 5.7.4a)
- Traceability between software requirements and tests (or other verification) is recorded (Clause 5.7.4b)
- The test results meet the required pass/fail criteria (Clause 5.7.4c)

Note also that your software development plan is expected by IEC 62304 to either contain or reference your verification planning. The most convenient way to address this on a universal level may be to reference your overall approach to the verification of process deliverables in your SOP. The verification work here is a part of that overall activity.

5.7.3 **Suggested Synthesis**

As we have sidestepped ISO 13485 verification and validation for now, the synthesis here is simple: follow the requirements made by IEC 62304. You should, however, ensure that you meet the requirements of ISO 13485 for the conducting of D&D reviews (see Section 5.9.1) and that your verification activities here are compatible with the requirements for conducting D&D verification (see Section 5.9.2).

Table 5.9 provides a glance over the main requirements of both standards. Note the software safety classification used by IEC 62304 and remember that some clauses of ISO 13485 may be excluded with a valid rationale (see Juuso 2022).

In other words, regardless of the software safety classification you are to establish tests, perform them, retest after changes, and link to software problem resolution (see Section 9) as appropriate.

5.8 Release

> *This is where we get ready to flip that plexiglas cover open and pull the trigger on our software to see the world, or to conversely wrap*

Table 5.9 Synthesis of Top-Level Requirements

Combined requirements	ISO 13485	IEC 62304 A	B	C
Establish tests for software requirements	-	✓	✓	✓
Use software problem resolution process	-	✓	✓	✓
Retest after changes	-	✓	✓	✓
Verify software system testing	-	✓	✓	✓
Software system test record contents	-	✓	✓	✓
Meet ISO 13485 requirements for verification review	✓	-	-	-

the software in some see-through plastic and ship it off. Whether it is your management representative, quality manager, or someone else punching the button you won't get to jettison the software without seeing to its support and maintenance from here on out. The limelight is on your product now, as illustrated in Figure 5.9.

IEC 62304 Clause 3.37 defines the term release as a particular version of a configuration item made available for a specific purpose. Clause 3.12 introduces the related term of "medical device software" which is noted to be a software system developed for the purpose of being incorporated into a medical device or intended for use as a medical device itself. Except for the possibility that a release is full or incremental (see Section 6.3.2) the two terms are highly compatible for a software product and may also include documentation and data relevant to the software. Note that IEC 62304 Clause 3.11 also attempts to define the term "medical device", but don't take this as a definition to override any regulatory requirements or definitions (see Section 1.3).

The relevant clauses of both standards are:

- ISO 13485 Clause 7.3.4 Design and development outputs
- ISO 13485 Clause 7.3.6 Design and development verification
- ISO 13485 Clause 7.3.7 Design and development validation
- ISO 13485 Clause 7.3.8 Design and development transfer
- IEC 62304 Clause 5.8 Software release

In the following, we will look at the expectations of both standards, one standard at a time, and then synthesize a plan to meet the requirements of both.

Figure 5.9 When the software is ready for the limelight, it's time.

5.8.1 Expectations from ISO 13485

The requirements for product release itself are addressed in Clause 7.3.8 on D&D transfer, but before you get to go there you must also take care of a few additional considerations expected by ISO 13485 and not discussed by IEC 62304. These stages cover the review of your D&D output, the wrap-up of your verification activities, and the validation of your medical device. Each of these steps is addressed in the following.

5.8.1.1 D&D Output Review

Earlier we handed the lead of our software-development activities to the IEC 62304 standard after discussing the D&D input stage (ISO 13485 Clause 7.3.3) that we saw as corresponding to the software requirements analysis stage in that standard. Therefore, it should be quite natural that once the development work is essentially done, we return to check what requirements the ISO 13485 standard has for the D&D output stage. The appropriate clause to return to is Clause 7.3.4 where the following is required of the D&D outputs.

- They meet the requirements set in D&D inputs (Clause 7.3.4a), i.e., we have those deliverables available we expected to have available after development.
- They are recorded in a form suitable for verification against the D&D inputs (Clause 7.3.4), i.e., the deliverables are recognizable in light of the requirements we had for them.
- They provide adequate information for purchasing, production, and service provision (e.g., distribution, installation, and maintenance) (Clause 7.3.4b), i.e., we know what it takes to manufacture and deliver our product.
- They instruct product acceptance criteria (Clause 7.3.4c), i.e., we believe we know what an acceptable product looks like and how to tell it apart from faulty products after the manufacturing line.
- They specify the essential characteristics for safe and proper use of the product (Clause 7.3.4d), i.e., we know how the product is to be used and we are ready to share this knowledge.
- They are approved prior to release (Clause 7.3.4), i.e., the deliverables are approved to be used for the next stages before moving on.

The activity here should be straightforward to perform if your software-development process has proceeded more or less as expected. All that is required here is that you compare your outputs against the inputs, record your assessment, and approve the outputs if they can be approved.

The inputs you are comparing against may be objects such as requirements, specifications, and perhaps design drawings. The corresponding outputs may then be, for example, lists of selected parts, assembly instructions, comprehensive instructions for the production line, and user instructions. In the case of a software product, the outputs may include the software binary, the installation package, and the instructions for use, for example.

In practice, the approval of the outputs may take the form of a design output review documented in your QMS (see Section 5.9.1 for documenting reviews). The review may reference various documents and the reviews conducted earlier during the software-development process. Here the objective is to ensure the outputs are fit for use by the later stages, i.e., to ensure everything that was expected to be produced has been produced. The correctness of the outputs will then be addressed in the verification stage up next.

5.8.1.2 Verification

The bulk of the work on performing D&D verification activities is discussed in Section 5.9.2. You have already looked at this section earlier when performing the verification steps expected in the previous stages by IEC 62304. ISO 13485 did not loom large in the earlier work. Here, before product release, the standard expects that your overall verification activities have been performed in accordance with your documented arrangements to ensure the D&D outputs meet the D&D inputs (Clause 7.3.6). This much is achieved by the output review discussed in the previous section.

In addition, in a closer study of the ISO 13485 standard it becomes apparent that it also wants to see the following verified:

- D&D outputs are verified as suitable for manufacturing (Clause 7.3.8)
- Purchased product that affects your product realization meets the purchasing requirements, and this must be based on supplier evaluation results and proportionate to the risks associated with the purchased product (Clause 7.4.3)
- Acceptance criteria for verification of medical-device installation activities must exist (Clause 7.5.3)

Furthermore, ISO 13485 expects that you verify, both now as is necessary and in the future as needs emerge, the following:

- D&D changes are verified before implementation (this is to be understood as before the changes take effect and can have any effect outside of their development)
- Production and service provision must be controlled, and these records verified (Clause 7.5.1)
- Product must be monitored and measured to verify that product requirements are being met (Clause 8.2.6)
- Monitoring and measuring equipment may need to be verified at specified intervals or prior to use against measurement standards and in accordance with documented procedures (Clause 7.6)
- Any customer property provided to the organization for use or incorporation into the product must be verified (Clause 7.5.10)
- If rework is performed the product must be verified to meet applicable acceptance criteria and regulatory requirements (Clause 8.3.4)
- If servicing is a specified requirement, instructions on verifying that product requirements are met after servicing must exist (Clause 7.5.4)

If the body of your verification work performed thus far falls short on any of these aspects here is the place to correct the situation. If you are confident that the verification work done meets all these needs, then congratulations: you have finished your verification activities for now. You may document this assessment in a verification review of its own, but a leaner approach is to document it in the Design Transfer Review coming up in a short while.

Note that the verification of the above aspects and their acceptance criteria, if specific to the product, should be instructed in any planning of product realization (Clauses 7.1c and 7.3.2c). This information must be available from your software development plan, but the plan may reference it from your SOP or another relevant document. Remember also that you must document the interrelation of personnel who verify work affecting product quality and ensure their independence and authority necessary to perform these tasks (Clause 5.5.1) – these are best addressed in the quality manual of your QMS (see Juuso 2022).

5.8.1.3 Validation

The validation work required for your product is addressed in Section 5.9.3. In addition to requirements discussed in that section for the product itself, ISO 13485 also wants to see that any production and service-provision processes where the output can't be or is not verified before the product is in use or the service has been delivered must be validated (Clause 7.5.6). This is primarily a topic for your QMS, but it is something worth pointing out here as we near the launch of a product developed under that QMS.

Also note that if your validation work causes any changes to be made to your earlier verification work, risk-management work, or other aspects of your product and its documentation you should react accordingly here. If all is as it should be you can move ahead to the release stage.

5.8.1.4 Production and Service Provision

Clause 7.5 provides a further list of items to consider before shipping the product. Here you are required to address the following:

- Control of production and service provision
- Cleanliness of product (incl. sterile medical devices)
- Validation of processes for production and service provision (incl. sterilization)

- Preservation of product
- Control of monitoring and measuring equipment
- Installation and servicing activities
- Identification and traceability
- Handling customer property

If you follow the QMS model set out in our earlier book (Juuso 2022), these will be covered by SOPs on infrastructure (SOP-3), suppliers and distributors (SOP-5), communications, sales, and post-market (SOP-7), and product realization (SOP-10). Your objective here is to consider the relevant aspects and make the provisions needed to ensure the right information and processes are in place to move forward with the launch. See Juuso (2022) for more discussion.

5.8.1.5. Release

After the D&D outputs and all verification and validation work has been approved it's time to move on to software release together with the IEC 62304 standard. The requirements for this leg of the journey in ISO 13485 are in Clause 7.3.8 on D&D transfer.

Here the fundamental requirement is that your organization has procedures for moving from D&D outputs to manufacturing. The principal requirement is now that D&D outputs are verified as suitable for manufacturing before coming to final production specifications, and that production capability is ensured to be capable of meeting the product requirements (Clause 7.3.8). As always, the results and conclusions of the transfer must be recorded.

In practice, it should be an efficient approach to develop a Design Transfer Review that accomplishes the above. This same document may also conveniently record the outcome of all verification and validation activities as it needs to comment on their outcome to provide clearance for moving ahead to production.

5.8.2 Expectations from IEC 62304

5.8.2.1 D&D Output Review

IEC 62304 is oblivious to the existence of any D&D output stage as understood by ISO 13485. Whether IEC 62304 is living in the Matrix, The Truman Show, or just plain doesn't care about such a mile marker on the way to a finished product is immaterial: only the ISO 13485 standard places requirements here.

5.8.2.2 Verification

Clause 5.8 for software release does not talk of any further verification activities, but it does require that all verification activity is complete and that all results have been evaluated. It therefore makes sense to now take stock of everything we have done. As a reminder, you have already verified the following according to the IEC 62304 standard:

- Software requirements (Section 5.2)
- Software architecture [B C] (Section 5.3)
- Detailed design [C] (Section 5.4)
- Implementation of software units [B C] (Section 5.5), and that any test procedures involved have been assessed to be adequate
- Software integration [B C] (Section 5.6), and that any test procedures involved have been assessed to be adequate
- Software-system testing (Section 5.7), and that the appropriateness of the verification strategies and test procedures is evaluated

In addition, the IEC 62304 requires that you verify, both now and in the future as needs emerge, the following:

- The effectiveness of any changes made to the software (Section 5.7)
- Documented risk-control measures have been verified [B C], and this includes an assessment of whether the measures may lead to new hazardous situations (Section 7)
- Resolved problems have been verified (Section 9)

In practice, the completion of all verification activities should also be checked and stated in some review documents. Thus, you should verify that you have verified everything you needed to verify. The practical place to do this is addressed below along with the discussion on IEC 62304 expectations on release.

5.8.2.3 Validation

The earlier D&D output review was something IEC 62304 was clueless about, but the same can't be said of validation. Here IEC 62304 is adamant that it does not cover validation, not even when the software developed constitutes the entirety of the medical device being released. The standard offers up IEC 60601-1 and IEC 82304-1 as possible sources of instruction on validation activities (see Section 12.2.3).

The standard does want to see the manufacturer synchronize software development with the development of the wider system, including the validation of that system, in some way in the software development plan (Clause 5.1.3b). In other words, IEC 62304 knows there are perimeter fences around it but rather than poke through them it is happy to graze on and let other standards govern those outer parts of the world. If only the raptors in *Jurassic Park* had been as complacent the humans on the other side of the fences could have been saved a great deal of harm and hazardous situations.

5.8.2.4 Production and Service Provision

IEC 62304 does not approach production and service provision explicitly. The standard does expect processes for software maintenance, change management, and problem resolution to be in place and functioning appropriately. It also speaks of post-production information and a necessity to maintain risk-management activities in a few contexts and thus expects these to be up and running for the life cycle of the device.

5.8.2.5 Release

Finally, Clause 5.8 sets the following requirements on the manufacturer for software release to take place.

- **Completion of planned activities**
  Ensure that all software-verification activities are complete, and their results have been evaluated (Clause 5.8.1). In case of B- or C-class software, you are also to ensure all software development plan (or maintenance plan) activities and tasks are complete and the associated documentation created (Clause 5.8.6).
- **Establishment of delivery procedures**
  Ensure you have established procedures to establish reliable delivery of the released medical-device software to the point of use without corruption or unauthorized changes (Clause 5.8.8). The procedures must address the production and handling of media containing the software product (including, as appropriate, replication, media labeling, packaging, protection, storage, and delivery).

- **Documentation of residual anomalies**
  Ensure that all residual anomalies are documented (Clause 5.8.2), and in the case of B- or C-class software, also evaluated to ensure they do not contribute to an unacceptable risk (Clause 5.8.3).
- **Documentation of version**
  Ensure that the version of the medical-device software release is documented (Clause 5.8.4), and in the case of B- or C-class software, the documentation also addresses how the software was created (i.e., the used procedure and environment is documented) (Clause 5.8.5).
- **Archival of software, configuration items, and documentation**
  Ensure the medical-device software, configuration items, and documentation are archived (Clause 5.8.7a-b). The minimum time the archived items are to be available is defined as the longer of a) the lifetime of the medical device as defined by the manufacturer or b) the time specified by relevant regulatory requirements.

Note that many of the above requirements previously only applied to B- and C-class software, but since the 2015 amendment the large majority now also applies to class A software. The exceptions to this are noted above.

The IEC 62304 does not require any document by a specific name here, as indeed it doesn't do elsewhere either, but here no specific document is even implied to exist. In practice, recording the above assessments is necessary, and the most convenient location to do so in your documentation is discussed next in the suggested synthesis.

5.8.3 Suggested Synthesis

Table 5.10 provides a glance over the main requirements of both standards. Note the software safety classification used by IEC 62304 and remember that some clauses of ISO 13485 may be excluded with a valid rationale (see Juuso 2022).

In general, the two standards can be quite neatly combined by fitting most of IEC 62304 requirements between what is called the D&D input stage and D&D output stage of ISO 13485. The quality-management system standard is happy giving the larger framework for the work, and the software-development standard is only too happy to dive into the details of the actual development work. For the most part the requirements of ISO 13485 can be conveniently used to shade or augment the meaning of what a plan, a review, or verification activity means.

Table 5.10 Synthesis of Top-Level Requirements

Combined requirements	ISO 13485	IEC 62304		
		A	B	C
Ensure D&D outputs meet D&D input requirements	✓	-	-	-
Ensure software verification is complete	✓	✓	✓	✓
Ensure software validation is complete	✓	-	-	-
Ensure D&D transfer requirements are met	✓	-	-	-
Document known residual anomalies	-	✓	✓	✓
Evaluate known residual anomalies	-		✓	✓
Document released versions	✓	✓	✓	✓
Document how released software was created	✓		✓	✓
Ensure activities and tasks are complete	✓		✓	✓
Archive software	✓	✓	✓	✓
Assure reliable delivery of released software	✓	✓	✓	✓

There are, however, additional profoundly important requirements on what the ISO 13485 wants to see take place before that thing you made is released to the world. The microscope here is focused squarely on the medical-device product you will make, and not the software units, items, and systems IEC 62304 had its eye on. The work IEC 62304 had you do is profoundly important, too, none of it will be wasted here, but as that standard was quite clear from the start that it didn't address the validation or final release of a medical device (see Clause 1.2 under Section 3.2), here is where ISO 13485 steps in.

The earlier software development stages have already seen to most of the heavy lifting involved in developing and testing your software. The steps required by ISO 13485 on top of any requirements made by IEC 62304 stages are the D&D output review, D&D verification, and D&D validation. These will be addressed in the following via the Design Output Review (DOR), the Design Transfer Review (DTR), and the validation report.

5.8.3.1 Design Output Review (DOR)

As noted above, the Design Output Review is where you will take stock of the results produced by your development activities. Here in the release stage, IEC 62304 sees all your verification activities as finished and it is time to check that all the requirements placed on the activities by ISO 13485 are also satisfied. In essence, what is required here is that you look at your IEC 62304 verification results from the earlier development stages and assess whether these together meet the requirements of ISO 13485. See Section 5.9.2 for details.

By working on the implementation of your software, you will have already performed several verification reviews that can now be referenced here. These verification steps include both software-unit verification (see Section 5.5) and software-integration verification (see Section 5.6) thereby addressing both the trees and the forest of your software as IEC 62304 sees it. You have also evaluated the adequacy of your verification strategies and test procedures and checked that all software requirements have been verified when performing software-system testing (see Section 5.7 for both). Here you will want to make use of these reviews instead of repeating work. Refer to Section 5.8.1 for the contents of the review. Also remember to check on the verification requirements from IEC 62304 detailed above in Section 5.8.2, naturally.

If everything has gone smoothly you are probably quite happy drawing up the review here. If all is well, you can close the review and move to performing or perhaps finishing your validation work.

5.8.3.2 Validation Report

Taking the waterfall model of software development, your validation activity will take place after the verification activity has been completed to the desired effect. In principle the verification and validation activities (often just called V&V) may be completed piecewise and may overlap, but as validation may involve costly clinical studies you will want to make sure your software works as you intended before sending it out for validation. In practice, your clinical evaluation report (CER) will be a big piece of the puzzle here. Validation is discussed in detail in Section 5.9.3.

After validation is in check you can move on to performing your third big review, the Design Transfer Review.

5.8.3.3 Design Transfer Review (DTR)

The last of your three big software development reviews is the Design Transfer Review (DTR). Here the objective is to complete any final assessments for the release to take place as described above. You may also take this opportunity to once more go over your product's risk-management activities to see if the validation work you performed should have any impact on those activities.

A key factor here is that if you have defined product acceptance criteria in your SOP or in the product-specific documentation (see DOR above) those are satisfactorily met here. The ISO 13485 standard speaks of product-acceptance criteria (Clauses 7.1, 7.3.4, and 8.2.6) in the sense that faulty products coming from your production line can be detected, but as software production is usually akin to cloning after the development is complete the place to ensure product acceptance is here when making the transfer to production. Therefore, you should ensure that any product-specific acceptance criteria from the DOR are assessed here, and you may also choose to look at all or some of the following.

- All work defined in the software development plan (SDP) is satisfactorily complete. This includes, in particular, any testing, verification, and validation activities.
- Product documentation meets the expectations of IEC 62304 (e.g., DOR supports moving forward) and any regulatory requirements (e.g., TF, IFU, label, DoC, and UDI). Note that some final documents, such as the declaration of conformity (DoC) may not be signed yet.
- The product and its documentation have been developed under your QMS, and the status of that QMS is appropriate (e.g., ISO 13485 certification is valid)
- The risk-management file of the product meets the requirements of ISO 14971 and your QMS. The file is complete and its final benefit/risk assessment supports releasing the product.
- The information expected for product and service provision is in place (incl. installation, training, and servicing/maintenance) to the extent that is defined as appropriate by your D&D documentation. Note that this may have already been satisfactorily achieved by the DOR.
- The processes for product maintenance and post-market surveillance are in place.
- Regulatory approval/clearance has been conducted in accordance with the regulatory requirements or will be prior to the launch of the product or is not appropriate for this type of product.

- No obstacles, aside from any pending regulatory approvals/clearances, are seen to exist for releasing the product and signing the declaration of conformity

Note that ISO 13485 speaks of various types of acceptance criteria in many clauses (e.g., 7.3.6 for verification, 7.3.7 for validation, 7.4.2 for purchases, 7.5.3 for installation, 7.5.6 for process validation, 8.3.2 and 8.3.4 for fixed product), as does the IEC 62304 (Clause 5.5.3 for software units, and Clause 8.2.1 for changes), but these are not directly relevant to the topic of acceptance criteria for your final product. Both standards do, of course, expect both your product and processes to meet the requirements of the standards, the user, and any applicable regulatory requirements. The place you have to be satisfied this really is the case is here, but the heavy lifting involved in making the necessary checks has hopefully already been done when following your instructed processes.

In practice, you may want to look at the above list in detail when getting ready to release version 1.0.0.0 of your product but then employ a streamlined approach when making subsequent releases. The streamlined approach could focus on the changes made, the continued usability of your previous DTR or its revision, and any other developments of import. The exact chronological order of the DTR and your regulatory approvals/clearances may also be debatable, but generally speaking you should not sign a declaration of conformity before the product truly is fit for release. A typical marching order here might be that you conduct a DTR, use those documents for communicating with the regulatory body (incl. an unsigned draft of a DoC), and then sign the DoC once given a green light from your regulatory body.

Finally, after making sure any required regulatory approval activities have been favorably completed you may proceed to launch your product in accordance with your SOPs. This may entail passing a conformity assessment, obtaining a UDI ID (if not yet done), completing your device label, signing a declaration of conformity, and making registry entries (e.g., EUDAMED actor and device registrations in Europe). These are all activities you should start preparing for in advance, of course.

After all these steps are complete you may take a breath and let your distributors and customers gulp up the merchandise. You will, of course, already be thinking of the next release and you must be ready to react to run your post-market surveillance activities and react to any support requests and issues that pop up in the meantime. In any case, this is a good time to crack open a case of champagne.

5.9 The Parts Left Out by IEC 62304

5.9.1 Conducting Reviews in D&D Stages

For IEC 62304 Clauses 5.3 through 5.6, the ISO 13485 standard does not place requirements that talk directly to the activities, tasks, or deliverables produced. These are topics the IEC 62304 standard conveniently adds between the requirements of D&D inputs and D&D outputs as discussed in ISO 13485. However, in the case you will want to claim that these stages are performed and approved in conformance with the wider ISO 13485 standard you must also ensure that the reviews performed during these stages comply with the requirements this standard sets for D&D reviews in its Clause 7.3.5.

The chief requirement in Clause 7.3.5 is that you must perform systematic reviews at suitable stages throughout your D&D activities. The reviews must be performed in accordance with planned and documented arrangements (i.e., observing SOPs, software development plans, and any other arrangements made).

The objective of a review is to a) evaluate the ability of any result from the stage to meet the requirements, and b) identify and propose any necessary actions. The review must include participation from representatives of the D&D stage concerned, as well as other specialists as needed. You may also benefit from involving clinical expertise or an independent reviewer in various reviews. In terms of records from the review, a simple review document identifying the usual details of exactly what was reviewed, by whom, when, and to what effect is the basic expectation here. Remember also to comply with requirements for the control of records as defined by ISO 13485 in Clause 4.2.5. Your overall QMS will take care of the last mile here in terms of documentation control.

5.9.2 D&D Verification

Both ISO 13485 and IEC 62304 lean on ISO 9000 for the definition of the term verification and are thus in apparent agreement with what it means. Verification, in general, is the confirmation through provision of objective evidence that specified requirements have been met.

This may entail measurement, but in many cases, verification is essentially synonymous with a review of evidence. What is required here, from the point of view of ISO 13485, is that we assess the output in light of the input and determine whether what you got is what you wanted to get.

In other words, ISO 13485 sets the scene for how verification activities are to be performed, IEC 62304 then gives us more specific requirements to look at during some of those verification activities (as detailed in the previous sections of this book), and your documented arrangements should then tell us at what milestones each verification activity takes place. Remember that your software development plan is expected by IEC 62304 to either contain or reference your verification planning. The most convenient way to address this on a universal level may be to reference your overall approach to the verification of process deliverables in your SOP. The verification work here is a part of that overall activity.

Here, ISO 13485 Clause 7.3.6 builds on the concept of the D&D review introduced above to require that you also perform verification – in accordance with your SOP, plans, and other documented arrangements – of the outputs meeting their input requirements. Note that unlike D&D reviews, verification activities are not said to take place at suitable stages during the development work. Instead, D&D verification as set out in the standard sits as a stage of its own between the body of the development work and the release of that work. Verification, however, does not have to be a separate stage but can instead be built to make use of all earlier work, including that performed during the unit and integration phases presented by IEC 62304. This is also acknowledged by ISO 13485 Clause 7.3.2c which tasks D&D planning with identifying verification activities that are appropriate at each D&D stage. Naturally, if there are any further verification needs not met by these stages you will want to address those needs too before release (i.e., D&D Transfer in ISO 13485 parlance) takes place.

The standard does not require that you develop and describe a process for this in an SOP, only that you perform verification appropriately and follow any arrangements you have made. You must, however, have a plan for verification that addresses the methods, acceptance criteria, and (if appropriate) statistical techniques including a rationale for sample size. A special consideration here is that if your product is intended to be connected with another device or interface this must also be covered in your verification. As always, records will be kept, and these will include any conclusions made, and necessary actions identified.

In practice, verification done according to ISO 13485, or according to IEC 62304 for that matter, is largely an internal activity ensuring that what you obtained via the D&D process matches with what you set out to create. The objective here is to compare the D&D output and the D&D input and assess whether these match as expected. The result of verification may be conveniently documented as a review or a report summarizing the results and approving the outcome. Here the overarching goal is to ensure you are content with the output so far and feel confident to proceed to the next stage. If you are not, you should return to a previous step as is appropriate.

IEC 62304, as mentioned previously, sees verification as taking place as part of the individual software-development stages. It is up to you how you marry these expectations but performing piece-wise verification throughout the D&D stages as expected by IEC 62304, making sure this verification is compatible with the above requirements of ISO 13485, and then pulling the whole body of verification work together before product release is a good plan.

5.9.3 D&D Validation

The definition for validation used by ISO 13485 is derived from ISO 9000 Clause 3.8.13 where the term is defined as confirmation based on objective evidence that the requirements for a specific intended use or application have been fulfilled. The point of comparison is thus the intended use (intended purpose) of your medical device.

Similar to D&D verification, D&D validation sits as a stage of its own preceding the release of the developed product (Clause 7.3.7). Unlike verification, though, validation is expected to take place using representative product (e.g., initial production units) and the work is expected to ensure the product is capable of meeting the requirements for its specified application or intended use. The rationale for choosing the representative product to use here must also be recorded. As with verification, here too you must observe connections to other devices and interfaces implicated in the intended purpose (intended use).

ISO 13485 Clause 7.3.2c does open the door to performing validation activities throughout the D&D stages by tasking D&D planning with identifying validation activities that are appropriate at each D&D stage. This is not mentioned by the main clause on the subject, Clause 7.3.7, and may be difficult to implement given the above requirements, but it is thus possible. If the above requirements can be met for some portion of the final device, and the validation work thus carried out in smaller meaningful segments, the overall validation work may then be tied together at the end – but you may run into issues addressing the interplay of the parts and the entire device they comprise.

Furthermore, validation is to include clinical evaluation or performance evaluation of the product in accordance with applicable regulatory requirements. The two terms are defined in Clauses 3.3 and 3.17 as follows.

- **Clinical evaluation (for medical devices)**
  Referring to the verification of the clinical safety and performance of a medical device when used as intended by the manufacturer (Clause 3.3), this is based on the assessment and analysis of clinical data pertaining to that device.
- **Performance evaluation (for in-vitro diagnostic devices)**
  Referring to establishing or verifying the ability of a device to achieve its intended use (Clause 3.13), this is based on the assessment and analysis of some data.

Since the requirement for validation is from ISO 13485, and since IEC 62304 sits on its hands when it comes to validation, this is something required of all medical-device software regardless of its software safety classification, for example. In practice, the general risk-based approach, guiding virtually all approaches to standardization and regulation, and the device classification for your product (e.g., I–III) will dictate what is expected of your clinical or performance evaluation here. You may find that performing a survey of literature and public databases of devices, adverse events, and clinical trials may get you far, but you may also discover that as the risk profile goes up the larger the clinical studies you will have to perform get. You can, of course, start work on many of these information sources before your final device is ready.

In any case, this is a stage where you will again benefit from listening to your customer. Understanding what your customers, users, and patients expect and what they think of your product may be the only way of knowing that you all want to see the same species of trees as illustrated in Section 1.14. You may want to consider turning to the various testbeds run by hospitals, medical centers, and other operators in your area to ensure an appropriate involvement of representative end users.

Remember also that you must have a validation plan that, like a verification plan, addresses the methods, acceptance criteria, and (if appropriate) statistical techniques including a rationale for sample size. All of your validation work must also be in accordance with your overall planned and documented arrangements. Your validation activities must then be completed prior to the release of the product, but to avoid chasing your tail note that a device used in the context of validation is not considered as released for use to the customer.

Finally, records of the results, conclusions, and any necessary actions must naturally be kept. In practice, a validation review or a report is a document you will want to generate here as you will be referring to that document in many different places and also for a long time to come. Also, IEC 82304-1 on health software expects both a validation plan and a report to be established and sets requirements for both (see Section 12.2.3).

The activities in validation may be a vast exercise depending on your product and affinity with clinical work, or it may be quite straightforward to pull together building on from your existing customer processes, but it is nonetheless a required part of manufacturing medical devices. Note that due to the regulatory requirements on post-market clinical follow-up and post-market surveillance this will not be a once-off thing either; instead you will be monitoring and building on the clinical evaluation over the years to come.

Chapter 6

SOFTWARE MAINTENANCE

This is the part that's easy to neglect when first sketching out some revolutionary new product on a napkin, but maintenance is a key part of the proposition that allows your customers to reap the benefits you envision, and for you to remain in business. It is also required – not merely to maintain your products, but to also keep an eye over them and churn out valuable post-market surveillance information. In cinematic terms, it may not be enough to leave an oblivious Tom Cruise behind on Earth to oversee those gigantic water pumps – at least plan to call him up every once in a while to see if he still thinks he is effective at maintenance, as indeed happens in the film.

Depending on your business model, you may be counting on maintenance to bring in much of the revenue, or you may be hoping that nothing really pops up for you to repair. In any case, you should not just be waiting with a toy mallet for those monsters to pop out so you can squish them in record time like in that carnival game. You should streamline your maintenance processes so that they are effective to execute if and when you find an urgent patch or a security update is needed. You must also keep an eye open and use both active and passive methods to predict, pinpoint, and fix any issues from here to eternity – or whatever you have defined as the supported lifetime of your device.

The topics covered in this clause are given in Table 6.1. Note that all the requirements of Clause 6 apply to all software safety classes. The table also adds a final section on the aspects not covered by IEC 62304 but expected by ISO 13485.

Table 6.1   Topics of Software Maintenance

#	Topic	A	B	C
6.1	Software maintenance plan	☑	☑	☑
6.2	Problem and modification analysis	☑	☑	☑
6.3	Modification implementation	☑	☑	☑
6.4	The parts left out by IEC 62304	☑	☑	☑

☑ = all apply, ☒ = some apply, ☐ = none apply

In general, ISO 13485 doesn't discuss maintenance for the product in any great detail even though maintenance and long-term support is part of its overall worldview on medical devices. It expects your organization to determine requirements related to, for example, delivery and post-delivery activities, and the subsequent maintenance activities to take place, be documented, and even be a source for dredging up silent customer complaints (see SOP-7, SOP-10 in Juuso 2022). Its Clause 0.1 even goes as far as to suggest that the standard as a whole is also applicable to maintenance organizations, implying that many of the processes discussed may in fact be invoked in the context of maintenance. IEC 62304, on the other hand, adds a great detail of instruction on the maintenance of medical-device software.

6.1 Software Maintenance Plan

It is not enough to build it and let them come. You should also have mechanisms in place for handling the issues and ideas for improvement that may pop up once the first version of your product has sailed. You don't have to have that "James Bond will return in X" signoff at the end of your credits, but you have to have an idea, a process, and hopefully, all the necessary resources to run that maintenance process for the foreseeable future. Figure 6.1 illustrates this.

Software maintenance is the flip side of the coin when it comes to software development. It is the afterlife of your development project, but at the same time an integral part of manufacturing and releasing medical-device software. The topic is addressed by both standards with compatible views but different points of observation. ISO 13485 looks at everything from the point of the status quo, your overall quality-management apparatus, and the users

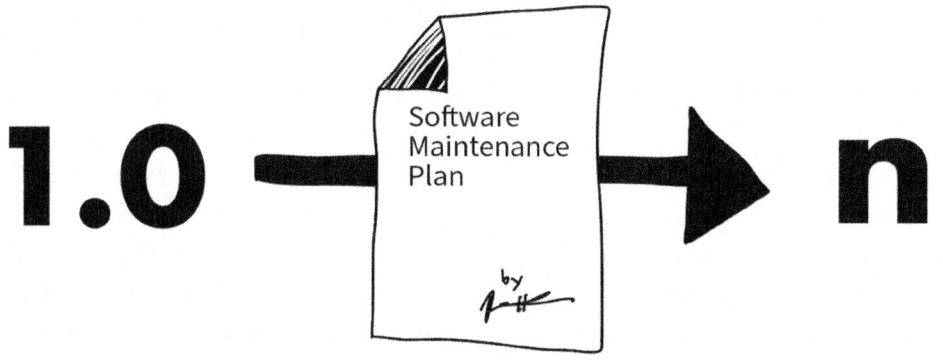

Figure 6.1 Software maintenance must be on your roadmap.

in mind. IEC 62304 appreciates all of this but takes a more grassroots-level approach to developing a plan for maintenance and then executing on elements of that plan almost as needed.

The relevant clauses of both standards are:

- ISO 13485 Clause 7.2.1a Determination of requirements related to product
- ISO 13485 Clause 7.2.3 Communication
- ISO 13485 Clause 7.5.1 Control of production and service provision
- ISO 13485 Clause 7.5.4. Servicing activities
- ISO 13485 Clause 7.3.9 Control of design and development changes
- IEC 62304 Clause 6.1 Establish software maintenance plan

In the following, we will look at the expectations of both standards, one standard at a time, and then synthesize a plan to meet the requirements of both.

6.1.1 Expectations from ISO 13485

The ISO 13485 doesn't have a clause on a software maintenance plan, no such document is in fact required by the standard, but a manufacturer is required to have a documented process for controlling D&D changes (Clause 7.3.9). Technically controlling changes is not the same as being equipped to do maintenance and actively staying on top of maintenance needs, but in practice, the two are coupled. The requirement to monitor both production and post-production information and act accordingly is implicated by the introduction of the term post-market surveillance (Clause 3.14) and stated, for example, in Clause 7.2.1 where you are expected to determine the requirements related to the product including any post-delivery activities. Clause 7.5.1f then requires you to implement the determined activities. Similarly, Clause 8.2.1 requires you to gather information from both production and post-production activities, and any issues detected here should not end up under the carpet.

Also note that Clause 7.5.4 directly discusses servicing activities related to the product. Although the frame of thought here is first and foremost on physical products shipped to the customer and requiring maintenance to stay on-spec it may be worth comparing your situation to the requirements. What is required is that you determine the servicing activities required for your product and then define the processes and reference materials as relevant. You are also required to keep records of maintenance activities regardless of whether these are performed by you or your supplier and analyze the records for possible complaints. In the case of a SaaS product, you may find that continued maintenance will also involve assessing updates and revalidation needs of the tools and SOUP components your product may utilize.

Similar to Clause 7.5.4, also note that as a small but potentially quite important detail, Clause 6.3 discusses the need to document maintenance activities for infrastructure, which may or may not be relevant here depending on the role of that infrastructure in the maintenance of your product and its quality. You are, of course, required to maintain the ability of your entire QMS to allow your processes to run as intended, and ensure resources and competencies across your organization as needed. Given this is the case, the link to your product may be more direct than is obvious at a first glance to Clause 6.3, and thus you may want to make the appropriate provisions in your software maintenance plan. If, for example, you maintain a server backend or a cloud-based software system it may be critical to address the maintenance of this part of your product adequately here. In the same vein, note that Clause 7.5.1 sees control of production and service provision as a continuous practice to ensure the product conforms to its specifications. Here too the emphasis is on the production environment instead of the product out in the field, but for a software product the two may have a strongly symbiotic relationship.

In addition, you will want to address how the various types of problems discussed in Section 9 will be linked to your software-maintenance process. For example, how customer feedback and complaints, detected nonconforming products or other nonconformities may factor in. Customer communication (Clause 7.2.3) in particular is a topic invoked by IEC 62304 in Clause 6.1, but also addressed in depth by ISO 13485 in a more general context (see SOP-7 in Juuso 2022). The role of rework may be something you wish to address or exclude in the plan or better yet in the higher-level SOP (see SOP-10 in Juuso 2022).

6.1.2 Expectations from IEC 62304

The manufacturer is required to establish one or more maintenance plans for conducting the tasks and activities of Clause 6. The plan must address the following:

- Procedures for receiving, documenting, evaluating, resolving, and tracking feedback arising after the release of medical-device software (Clause 6.1a)
- Criteria for determining if the feedback constitutes a problem (Clause 6.1b)
- Use of the software risk-management process (Clause 6.1c; see Section 7)
- Use of the software problem-resolution process for analyzing and resolving problems arising after the release (Clause 6.1d; see Section 9)
- Use of the software configuration-management process for managing modifications to the existing software system (Clause 6.1e; see Section 8). Note that here the term "software system" is used instead of the term "medical device software" used above
- Procedures to evaluate and implement upgrades, bug fixes, patches, and obsolescence of SOUP components (Clause 6.1f)

Also keep in mind that IEC 62304 expects to see a maintenance plan covering work according to the maintenance requirements it makes in Clause 6, not just matching the above list.

6.1.3 Suggested Synthesis

The requirements of the two standards are complementary through and through. Table 6.2 provides a glance over the main requirements of both standards. Note the software safety classification used by IEC 62304 and remember that some clauses of ISO 13485 may be excluded with a valid rationale (see Juuso 2022).

Table 6.2 Synthesis of Top-Level Requirements

Combined requirements	ISO 13485	IEC 62304		
		A	B	C
Documented process for controlling D&D changes	✓	-	-	-
Documented process for feedback handling	✓	✓	✓	✓
Documented process for implementing upgrades, bug fixes, patches, and obsolescence of SOUP components	-	✓	✓	✓
Criteria for feedback constituting a problem (complaint)	✓	✓	✓	✓
Determine servicing activities	✓			
Use of risk management to advise planning	-	✓	✓	✓
Use of problem-resolution process	-	✓	✓	✓
Use of configuration management	-	✓	✓	✓

Both standards expect a range of processes to be in place, although you are perhaps allowed to define your own relationship with activities such as servicing and control of the production environment to the extent it corresponds with the platform for deployment. Remember also that ISO 13485 lets you exclude certain clauses with a valid rationale, but this should be done on a high level in your quality manual or SOPs, not in your maintenance plan (see Juuso 2022).

In fact, all the necessary instructions could be contained in an appropriate SOP and not your maintenance plan. In practice, though, you will want to make product- or project-specific arrangements and resource allocations, and a good place to document these may be in the maintenance plan. The process for the actual change control is addressed in Section 8.2, but here you are expected to address the possibility of changes and the overall process for maintenance work.

6.2 Problem and Modification Analysis

Problem and modification analysis sounds like something Georg Clooney's suave character in Up in the Air *might be into, both professionally and in his personal life. This is where you land at the realization that something might need to change, and you consider all of the effects that change might have. This is not yet where you run to Vera Farmiga's front door and hope to give your life a romantic root canal. Figure 6.2 illustrates another example of begging for a reaction.*

Analyzing the need and the effects of change before diving into making those changes in your software product is something both standards can relate to. In general, software changes should not involve any amount of MacGyvering. IEC 62304 goes out of its way to instruct the change-management process here (see also Sections 6.3, 6.2 and 8.2), and ISO 13485 also has expectations on the matter. A good change-management process with an adequate analysis of the proposed change will be your key to figuring out if what you are embarking on really is a quick fix or something closer to a brand-new product development project. The changes you have made to your product, and how you have validated these changes (see Section 12.2.3), are a hot topic for future conformity audits – and topics where you don't want to have to start second-guessing yourself when seated in the hot seat.

The relevant clauses of both standards are:

- ISO 13485 Clause 7.3.9 Control of D&D changes
- IEC 62304 Clause 6.2 Problem and modification analysis

Figure 6.2   My dog ate my software.

In the following, we will look at the expectations of both standards, one standard at a time, and then synthesize a plan to meet the requirements of both.

6.2.1 Expectations from ISO 13485

The clause of ISO 13485 of most immediate concern to the analysis of modifications to the product through the provision of new releases is Clause 7.3.9 on the control of D&D changes. The topic is discussed in detail in our previous book (Juuso 2022), where SOP-10 contains the relevant guidance.

In terms of analyzing modifications, ISO 13485 calls for a proactive approach where each contemplated change must be identified, reviewed, verified, and validated (as appropriate) before implementation. Here verification is to be understood as verification before the changes take effect and can have any effect outside of their development.

Your organization is expected to determine the significance of the proposed change to the function, usability, safety, and applicable regulatory requirements for the medical device and its intended use. The review must also look at the effect on constituent parts, products in the process or already delivered, risk management (inputs and outputs), and the product-realization processes. Records, as always, must be kept on the reviews and any actions found necessary.

6.2.2 Expectations from IEC 62304

Clause 6.2 addresses your approach to problem and modification analysis, i.e., analysis of the need for changes. The fundamental requirements here are to monitor feedback, evaluate it, detect problems, use the problem-resolution process (see Section 9), employ change requests, and communicate with users and regulators as needed. Each of these requirements forms a brief subclause under Clause 6.2, which may not be ideal for readability as we have headings up to level four, but the requirements themselves are simple. Incidentally, a change request as a term is set in Clause 3.4 by defining it as a documented specification of a change to be made in the medical-device software. Thus, the term is not necessarily invoked for software that is not a part of the medical device.

Clause 6.2.1 requires the manufacturer to monitor feedback on medical-device software released for a specific intended use. The wording here raises questions about the possibility of filtering the feedback based on the defined intended use and leaves open your coverage of, for example, off-label use. You must document the feedback and evaluate it for possible problems in the product, though. If such problems are found, you are to record a problem report with actual or potential adverse events and deviations from the specifications (see Section 9). Finally, you are to evaluate the reports to determine whether product safety is affected (considering the intended use also) and whether a change in the product is needed.

Clause 6.2.2 drives home the fact that you are to use your problem-resolution process to address the problem reports. A nonbinding note highlights the possibility that the problem could show that the software safety classification of a software system or a software item may be incorrect. The problem-resolution process may therefore suggest changes to the classification, and if so, this is to be documented and communicated after the completion of the resolution process.

Clause 6.2.3 calls you to analyze change requests (which may follow from the analysis of problem reports) to assess the effect of the change on your organization, the product (considering the intended use also), and the systems it interfaces with. The object of it is left somewhat vague as it could refer to interfaces between the change request and systems, or the product and systems. Prior to the 2015 amendment this clause only applied to software safety classes B and C, but since the amendment the clause applies to all safety classes.

Clause 6.2.4 requires you to evaluate and approve change requests which modify a released product. The wording of the requirement leaves it open to interpretation whether what is meant are just changes when the modifications they make are released (at the point of release) or any changes to a working copy where an as-of-yet-unreleased new version of a software is modified. What is certain is that implementing change requests without evaluation and approval will be a red flag for conformity and for the smart use of your resources.

Clause 6.2.5 addresses the important topic of communicating with your users and regulators. Here you are required to identify the change requests that affect a released product. You must, as required by local regulation, inform your users and regulators about:

a) Any problem in a released product and the consequences of continued use unchanged
b) The nature of any available changes to the released product, and how to obtain and install these changes

In practice, Clause 6.2.5a shoots itself in the head by requiring notification of all detected problems. If this clause were to be followed the users and the regulators would be inundated by benign reports of minor feature and design changes made in response to detected problems. The recipients would not have sufficient resources available to address and understand what they are receiving, and far more serious problems would go unnoticed under the barrage. The goal is a good one, of course, but some triage is probably going to be called for in following this requirement. Note though that the trigger for communication here is not the change request, but the problem, and the problem is not trip-wired to the problem report. Thus, in practice, you should be able to employ common sense here and develop a process that ensures the right level of reporting for transparency and safety to be preserved without shell shock at either end.

Also note that the topic of Clause 6.2 is problem and modification analysis, which ought to imply that changes are possible also without the occurrence of problems and that these ought not always lead to regulatory reporting. In this sense, the clause appears somewhat old-fashioned and less than optimally suited for making improvements to a product and ensuring the smartest pipeline for facilitating the validation and release of those improvements.

6.2.3 Suggested Synthesis

The requirements of the two standards are complementary through and through. Table 6.3 provides a glance over the main requirements of both

Table 6.3 Synthesis of Top-Level Requirements

Combined requirements	ISO 13485	IEC 62304		
		A	B	C
Monitor feedback on the product	✓	✓	✓	✓
Detect problems and use the problem resolution process		✓	✓	✓
Use problem reports (incl. determine effect on safety)		✓	✓	✓
Use change requests (incl. analyze potential changes, approve appropriately)		✓	✓	✓
Communicate with users and regulators (incl. problem, impact of continued use without fixes, availability of fixes)	✓	✓	✓	✓

standards. Note the software safety classification used by IEC 62304 and remember that some clauses of ISO 13485 may be excluded with a valid rationale (see Juuso 2022).

In terms of software-related problems and modification analysis calling for a D&D change to be made, the above gives a good overview of the requirements. In case your problem corresponds with a complaint, a nonconformity or a nonconforming product, you will want to refer to the related processes in your QMS, as noted above, to ensure all the appropriate requirements are observed.

6.3 Modification Implementation

The cartoonish molding chamber from Streetfighter *and the reanimation process from the* Universal Soldier *are for us the first images conjured up by the topic of modification analysis. The business of enacting changes on your software product should be much more routine and much less sci-fi than these examples, but you may find that designing the right process to meet your needs here requires some deep thought. Figure 6.3 illustrates what the son of Frankenstein might look like if you're not careful.*

Figure 6.3 Problem-resolution process may be akin to field surgery.

The topic of how to develop and release software updates is an object of great fascination. The most bureaucratic process might be to treat every new version as a completely new software-development project. Your staff, customer, and regulators might not thank you for this. It would take far too long and be far too costly to enact any changes to the software, and thus many useful modifications would languish in the backlog. The simplest process would be to just do agile and push new releases out at will, but this would not meet the requirements. The sensible approach is somewhere in between.

The relevant clauses of both standards are:

- ISO 13485 Clause 7.3.9 Control of D&D changes
- IEC 62304 Clause 6.3 Modification analysis

In the following, we will look at the expectations of both standards, one standard at a time, and then synthesize a plan to meet the requirements of both.

6.3.1 Expectations from ISO 13485

As ISO 13485 does not make any special concessions here regarding maintenance updates, the basic expectation is that all of its D&D requirements, as detailed in Clause 7.3, are met. This may be understood to mean the validation and release requirements discussed in Sections 5.8 and 5.9 are upheld. The method of verifying this really is the case and the medium for recording the assessment may vary, but it must take place for the maintenance release to go ahead.

Remember also that ISO 13485 calls for changes to be identified, reviewed, verified, and validated (as appropriate) before implementation (i.e., before the changes take effect and can have any effect outside of their development).

6.3.2 Expectations from IEC 62304

Clause 6.3.1 requires the manufacturer to use their established process to implement the modifications assessed and approved in the previous section (Clause 6.3.1). This includes the identification and implementation of any Clause 5 activities (see Section 5) that need to be repeated because of the modification. A nonbinding note refers to Clause 7.4 for requirements related to risk management of software changes.

Clause 6.3.2 then requires the manufacturer to follow their process for making releases as discussed in Section 5.8. A nonbinding note explains that such releases may now take place as a) a full release of the software system or b) an incremental batch comprising the changed software items and the necessary tools to apply the modifications to the software system.

6.3.3 Suggested Synthesis

The requirements of the two standards are complementary through and through. Table 6.4 provides a glance over the main requirements of both standards. Note the software safety classification used by IEC 62304 and remember that some clauses of ISO 13485 may be excluded with a valid rationale (see Juuso 2022).

So, how do you design a safe, streamlined process for conducting software maintenance and making the necessary releases? The internal maintenance process up to the point of the release is simple enough to instruct. Where the beads of sweat start to appear on your forehead is in figuring out what work from the previous stages (ISO 13485 D&D and IEC 62304 Software development) needs to be redone to move ahead with the release. The great news here is that it depends on what has changed in the new version of the product, and this is to be evaluated by you here.

Based on what has changed you can go through the technical documentation and assess, even one document at a time, whether that document still holds true or whether the document needs to be revised somehow. If you come to the conclusion that some document needs to be revised, you

Table 6.4 Synthesis of Top-Level Requirements

Combined requirements	ISO 13485	IEC 62304		
		A	B	C
Observe D&D requirements the same way you would for a first release	✓	-	-	-
Observe software development requirements the same way you would for a first release	-	✓	✓	✓
Repeat software development activities as needed	-	✓	✓	✓
Use modification process to make changes	-	✓	✓	✓
Use release process	✓	✓	✓	✓

will then be able to assess how much of the work going into that document needs to be redone. After having gone through your entire set of technical documentation you will have an excellent idea of what has changed and what needs to still change as a result of allowing the new release to go through. Make sure you do not compromise on the release conditions you may have set in your processes, and that you are happy with the validation status and risk management file also covering the new release.

Over time as you get more experience with running the maintenance process you will start to have a better idea of what changes affect what earlier work and documentation. You will then be able to pre-empt some work while working on a maintenance release and pay less attention to some areas of the documentation. Section 11 will give you our map of the essential technical documentation, and the relationships between the various documents. This discussion will hopefully help you in streamlining your underground map from a change need to a released update.

6.4 The Parts Left Out By IEC 62304

There is both not a lot and a great deal that IEC 62304 does not cover regarding product maintenance. ISO 13485 sets up the foundations of performing the work from top management responsibility, resource allocation, and personnel competence to listening to and reacting to various sources of information and the events arising out of them as triggers to maintenance and problem-resolution activities. These were all covered above, and are discussed in greater detail in our previous book on ISO 13485 quality management (Juuso 2022). We are thus happy to wrap up software maintenance here and move to a topic as integral to medical devices as quality management itself: risk management. Both of these topics form an undercurrent to every single clause of IEC 62304.

Chapter 7
RISK MANAGEMENT

We've all seen War Games, Terminator, *and* 2001. *We can imagine what an autonomous software system on a rampage might look like. We also remember the AI from* The Hitchhiker's Guide to the Galaxy *that humanity placed its hopes on for generations only to discover after all that time that the meaning of life was 42. With great expectations come great fears, and already for that reason alone it is to be expected that risk management features heavily in any discussion of medical-grade software.*

The deep end of the pool for medical-device risk management is instructed by the international ISO 14971 standard. The intention here in Clause 7 of the IEC 62304 standard is to make some further requirements to address the special nature of software-specific risks, and to ensure that the risk management for the software and any device using that software occurs in some synchronized manner. The foundation for the use of both standards is, of course, set by ISO 13485.

Safety is the operative word for all three standards. ISO 13485 grounds risk as something pertaining to the safety or performance requirements of the medical device or the meeting of applicable regulatory requirements (Clause 0.2). The standard then does a bit of a doubletake to define the applicable regulatory requirements as the requirements set for the quality management system and the safety or performance of the device. To complete the circle here, even that performance part is occasionally explained by adopters of the standard to be examined via the safety of the device (e.g., the ability of the device to consistently function in a safe way). The top-level takeaway that emerges here is that the safety of the device is always key, and an appropriate QMS is seen as a key part of getting there. IEC 62304 takes a compatible view by placing great emphasis on the people coming into touch with the device, and how any risks may affect the patient, the operator, and any other relevant people. The goal is a device that is consistently safe for people. ISO 14971 looks over the two other standards here and nods approvingly on the course taken, but expects its provisions to be the most important for going forward.

We will look at all the assertions made by IEC 62304 on risk management next, and comment on the requirements made by the other two underlying standards and regulations as relevant. Before that, it is prudent to briefly introduce the cast of the last remaining concepts IEC 62304 has deemed important enough to import from ISO 14971 between its covers. The main concepts were already introduced above in Section 1.23 as hazardous situations, hazards, harms, and risks. Here we add a few more concepts to round out the discussion offered by IEC 62304. The terms here relate to the process phases and the risk management file itself. The new terms to consider are as follows.

- **Risk management**
    This is the top-level umbrella term (Clause 3.19) used to discuss the systematic application of management policies and processes to analyzing, evaluating, and controlling risk.
- **Risk assessment** (incl. risk analysis and risk evaluation)
    Risk analysis (Clause 3.17) refers to the systematic use of available information to identify hazards, as they are relevant within the context of the intended use of the device (or its foreseeable misuse), and estimate the risk involved. As part of this, risk estimation (Clause 3.39) is used as the process to assign values to the probability of occurrence of harm and the severity of that harm. Finally, risk evaluation (Clause 3.40) refers to the process of comparing the estimated risk against criteria defined on the acceptability of risk. Together all these steps comprise risk assessment, which is a term used by ISO 14971 but not introduced by IEC 62304.
- **Risk control**
    Risk control (Clause 3.18) refers to the process where risk-control measures (mitigations) are defined to reduce or maintain risks within specified levels.
- **Evaluation of residual risk**
    The concept of residual risk (Clause 3.38) simply refers to the risk remaining after risk-control measures have been assigned. The residual

risk may be considered for each risk by itself and for the entire set of risks overall.
- **Risk management review**

    The concept is not introduced by IEC 62304, but this is an integral part of your risk-management activities where the overall status of the activities is assessed and, usually, a decision is made on the benefit–risk ratio of your product. The review might, for example, be conducted for product-specific risk management, process-specific risk management, or both.
- **Production and post-production activities**

    Post-market surveillance and the continuous monitoring of the risks related to your product, but also the benefit gained by using it, are going to be a part of your follow-up activities for as long as the product is expected to remain in use – which may be longer than just the period it is sold. ISO 14971 and other standards and technical reports will guide this work (see Juuso 2022), but here it is useful to know that risk management might never actually end.

Finally, the term risk management file is introduced in Clause 3.20 as a set of records and other documents that are produced by a risk-management process. The clause interestingly enough points out that these records may not necessarily be contiguous, which should be interpreted as records created on a varying granularity and focus over time, but not as containing gaps or omissions.

It may also be worthwhile to recall the definition of "safety" as free from unacceptable risk (see Section 1.25) and to look at the definition of the intuitively related term of "security". Clause 3.22 defines security as the protection of information and data so that they are protected from unauthorized reading and modification but retains support for authorized access. Safety thus refers to the safety of people, but security is understood from the point of view of information.

The topics covered in Clause 7 are given in Table 7.1 below. The table also notes which software safety classes each topic is applicable to, but please don't make the mistake of thinking that class A software is somehow exempt from risk management as it is not implicated for Clauses 7.1–7.3. The requirements from Clause 4 apply to all classes of software, and the expectations from regulations and other key standards will call for risk management done with rigor even if not always with the same amount of work as would be done for high-risk software.

Table 7.1 Topics of Risk Management

#	Topic	A	B	C
7.1	Analysis of software contributing to hazardous situations	☐	☑	☑
7.2	Risk-control measures	☐	☑	☑
7.3	Verification of risk-control measures	☐	☑	☑
7.4	Risk management of software changes	☒	☑	☑
7.5	The parts left out by IEC 62304	☑	☑	☑

☑ = all apply, ☒ = some apply, ☐ = none apply

The need to adopt the ISO 14971 standard as a basis for your risk-management activities is, for now, an obligatory part of adopting IEC 62304 as defined in Clause 4.2. The call to adopt ISO 13485 as the guide for quality management is stated in the preceding clause, Clauses 4.1. Both of these justly factor into how you do risk management in the context of medical-software development.

There is, however, a third clause that leans in heavily on the implementation of risk management in our context here. That is Clause 4.3 which ostensibly addresses only the assignment and use of the software safety classification. The stipulations made in that clause are, however, routinely used in all risk-management activities for medical software. This is fair to the extent that the ABC safety classification does sit in the general requirements section of IEC 62304 and does fundamentally curate how we choose which requirements of the standard, including those of Clause 7, to heed in the circumstances of a specific software item. Where the utility of some of these assertions begins to break down a little is in the actual ISO 14971 risk management around a device, as we will discuss next. For a discussion of Clause 4 itself, refer back to Section 4.

The most striking assertions made by Clause 4.3 that often bleed over to Clause 7 are as follows.

- **The software safety classification is conceptually associated with the risks involved in the use of that software.** All software is to be assigned a software safety class, which by default is C, indicating a high risk. This classification is conceptually linked with the risks related to the software, but no definite formula exists for mapping the A, B, and C classes to the usual ISO 14971 risk classifications (e.g., 5x5 matrix of risk levels). Yet, you need to heed the impact of the safety classification on your risk-management activities, as is evident from Sections 7.1 to 7.4 below.

- **All software will eventually fail.** This is a steadfast expectation in the field and one that even junior developers may be able to recite. It is a handy assertion to make when figuring out the safety classification for your software as it does not allow you to claim that software could not fail just because it was programmed not to. As a carryover to then performing the day-to-day risk management related to that software, it is less practical and often circumvented by using both P1 and P2 as explained in Section 4.3. The bottom line here is that while all software will eventually fail (P1 is 100%), this failure may or may not lead to harm (P2 is something between 0% and 100%). Today, even ISO 14971 talks of P1 and P2, and interprets these in agreement with the status quo discussed above. Thus, to challenge the virtue of P1 = 100% would be to run the risk of embarking on a quixotic campaign against the windmills. In practice, you will see P1 in many risk matrices but its role in the analysis may be debatable if it is always 100%.

- **IEC 62304 occasionally only looks at harm in relation to people.** ISO 14971 understands harm as always relating to people, property, or the environment but when assigning the IEC 62304 software safety classification only consideration of people is required (Clause 4.3a). That may be fine for assigning the software safety classification, but it is not enough for conducting your subsequent body of risk-management activities. Here, it would be unwise to forget about property and the environment in your full risk management. In fact, there are a few more considerations you should also observe in addition to the standards as we will discuss in Section 7.5 in a short while.

- **Risk controls should be external.** Clause 4.3a suggests that external risk controls may be used to lower an initial software safety classification. This may be mistakenly read as only risk controls external to any software should be used, which is not what the standard expects. Although only risk-control measures external to or independent from the software itself may be considered when assigning the software safety classification (Clause 4.3a), it is in fact all types of risk controls, both internal and external, that should be considered when mitigating risks. These can be anything from a try-catch expression used in the source code to an elaborate check made by another module on the output value provided by some function. Software-based risk controls thus add a welcome solution to the arsenal available to software developers

and mean that hardware, environmental conditions, and procedural instructions are not the only means of mitigating risks for software products. Being able to rely on the independence of the software item and the control (i.e., being able to prove segregation between the item and another item controlling it) is important, though.
- **Risk control may decrease both the probability and/or the severity of a risk.** Intuitively this makes a great deal of sense; after all, you should be able to improve on both the probability and the severity of something going wrong if you put your mind to it. The matter is occasionally debated, but the standard is quite clear on it in a nonbinding note to Figure 3 in Clause 4.3a. Note that the context here too is software safety classification.

The above assertions represent some of the key philosophical pain points of risk management and are no doubt debated endlessly on various expert forums. Expect the whisky to run out before a complete consensus is reached on any of these matters. The above discussion should, however, help you navigate your way to meeting the requirements in practice.

One final requirement worth noting here before we dive into the subclauses of Clause 7, is that Clause 5.2.4 requires the manufacturer to not only re-evaluate risk assessment when software requirements are first established but also later update this assessment as appropriate. This means, ideally, that any change of software requirements – such as through change management – should be accompanied by reconsideration of the risk assessment. Don't go overboard here and require an unnecessarily wide overhaul just in case, but instead assess the situation and act accordingly when making software changes.

In the following, we will look at the individual subclauses of Clause 7. Note that the first three subclauses only apply to software safety classifications B and C, and significant leeway is given to class A also in terms of the fourth subclause. ISO 14971 risk management, though, is required from all classes of software by Clause 4.2.

7.1 Analysis of Software Contributing to Hazardous Situations [B] [C]

IEC 62304 is not unhappy with the instruction given by its big brother on risk management, the ISO 14971 standard, but it does

Risk Management ■ 195

> *want some software-specific issues to be observed. Here in Clause 7.1, its goal is to have you thinking of the specific ways the nature of software might contribute to the occurrence of harm. This is as if Tony Soprano brought his Sicilian third cousin over from Italy and we would be charged with predicting how he will fit into the New Jersey family. The heavy in the equation will still be ISO 14971, but IEC 62304 will not be a wallflower either.*

IEC 62304 may have pushed the envelope when it was first published by turning a keen eye on the software that may run in medical devices, or as medical devices, but it did so by principally looking at how that piece of software was put together and against what specifications. There is nothing wrong with that, but in a world that has moved from F-14 Tomcats to F-35 stealth fighters, the standard seems more aged than Tom Cruise with the same passage of time.

Today, it is IEC 82304-1 (see Section 12.2) that points out that it may no longer be all that savvy to just look at your software and the hardware surrounding it. Instead, you should look at the platforms your software runs on and the stealthy cybersecurity threats coming at your software. Anytime your medical-device software is connected to a network the potential risks are catapulted to a new level. Manufacturers don't always pay adequate attention to what the addition of a network connection does to the risk profile of the device and the risks associated with its features. As a result, the risk analysis may be decidedly inadequate. Such risks can't be eliminated by turning a blind eye to IEC 82304-1 either, they will still need to be on your radar via IEC 62304 Clause 7.1.2 (see below). Figure 7.1 below illustrates how the risk profile of the same functionality may have an increased risk profile as you move from a standalone device to a device connected to a private network, and finally, a device relying on a network of services in the cloud. Your environment is relevant to your risk analysis whether or not your standard has a word for the exact type of your environment.

In practice, the analysis of software contributing to a hazardous situation requires an in-depth understanding of not only the clinical procedure the device is used in and basic software engineering, but also of the network environment the software operates in. The risks may be related to the safety or the security of data and the system, for example. ISO 14971 is applicable to both safety and security risks, but it needs experience, insight, and judgment to identify and appropriately manage those risks – and it may benefit

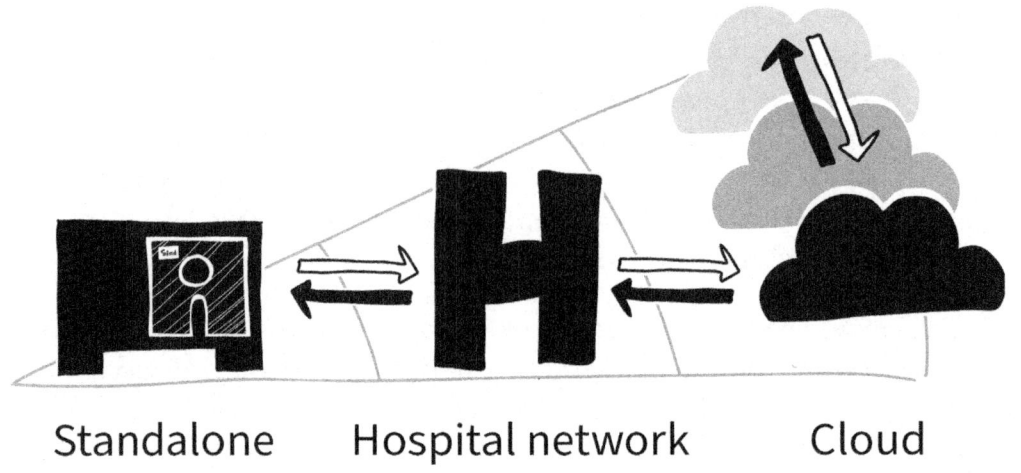

Figure 7.1 Standalone networks, private networks, and the cloud each come with different risks associated with them.

from the lens offered by IEC 82304-1 in this, too. Note also that as such new lenses are applied to identifying and managing risks, and as these are mitigated somehow, you may also end up affecting other risks or introducing new risks. These may lead to new requirements, new verification needs, and new test cases that then affect your overall design.

The use of a firewall may seem like a simple solution to countering many of the risks brought on by the addition of a network connection. That too, however, has more considerations than just a plug-and-play decision. You should, for example, consider the following.

- Do you treat the firewall as a SOUP component? If so, does the manufacturer of the firewall provide lists of known vulnerabilities? How are detected issues handled? How are updates and the need for updates managed?
- Have you identified all the risks related to the use of the firewall?
- Have you addressed all risks related to how your software communicates with the firewall?
- Does the use of the firewall place requirements on your software, e.g., in terms of how your software communicates with the firewall, or what security features it implements? Does this affect your requirements for any accompanying documentation?
- Does your validation adequately take the use of the firewall into account?

The above list is not exhaustive, neither is it intended to scare you away from using a firewall. The intention is only to point out that the situation is more complex when you have more components and more interfaces in play. This is also true when considering the interplay of devices, both medical and non-medical, cohabiting in the same network. It may, for example, transpire that some benign Zigbee devices intended for the home market but used in a hospital environment will be overeager to explore their surroundings and end up overwhelming their more limited hospital friends. That is a true story, and an example of how the network environment may mean that something thought too silly to happen previously (e.g., a janitor leaning against a light switch on the operating room wall) can now easily happen when the distance grows and the physical connection between the source and the point of the harm becomes more abstract (e.g., a janitor leaning against a wall switch in some remote control room). A filmic reference to the Michael Crichton thriller *Coma* comes to mind, but here the motive of the remotely controlled gadget does not have to be sinister. The example is intended to be illustrative of the unexpected complexity that may be introduced even when just taking previously available functionality into a new environment. It is up to you to identify, prioritize, and control the risks related to the use of your device.

7.1.1 Expectations from ISO 13485

As it happens, a similarly anointed Clause 7.1 in ISO 13485 expects the manufacturer to implement one or more processes for risk management in product realization, and thus an analysis of hazardous situations is required for medical devices in general. However, no specific requirements for software are made. The expectations for risk management in general are discussed in detail in our previous book (Juuso 2022).

7.1.2 Expectations from IEC 62304

For B- and C-class software, the manufacturer must identify software items that could contribute to an identified hazardous situation (Clause 7.1.1). Remember that software items may in fact be units, items, or systems following the definition of the term. In other words, you will have developed a risk management file for your device where you identified hazardous situations, and you are now required to pinpoint where in your software those failures might originate from. A non-binding note instructs you to consider not only software failures but also the failure of any software-based risk-control measures you have defined.

Clause 7.1.2 offers a helping hand in identifying such sources by instructing that you look at the following as potential causes for item failure:

- Incorrect/incomplete functionality specification
- Software defects in the functionality of the item
- Failure or unexpected results from a SOUP component
- Hardware failures (resulting in unpredictable operation)
- Software defects resulting in unpredictable operation
- Reasonably foreseeable misuse

The cumbersome list above could be summarized as the faulty specification for the item, faulty construction of the item, unexpected input to the item (from SOUP, hardware, or other software), and reasonably foreseeable misuse. In effect, the clause requires you to not only identify the suspect items but also consider the potential circumstances that could lead to the item contributing to a hazardous situation.

Clause 7.1.3 examines SOUP components more closely. Provided that the failure of a SOUP component, or the unexpected output of such a component, is a potential cause for an item to contribute to a hazardous situation, you are required to take action to control SOUP components. This must, at a minimum, entail the evaluation of any anomaly list published by the SOUP supplier for the particular SOUP version used in the medical device. The evaluation is to include a determination as to whether any of the anomalies could lead to a sequence of events potentially leading to a hazardous situation.

Finally, Clause 7.1.4 requires the potential causes behind an item contributing to a hazardous situation to be documented in the risk management file. Note that the old subclause 7.1.5 has been deleted from the amended version of the standard and you are no longer required to document sequences of events potentially leading to hazardous situations in the risk management file.

Please note that class A software gets a free pass in terms of clauses 7.1–7.3 discussed here, but it does not escape risk management altogether as detailed in Section 4.2 above and Section 7.3.1 below.

7.1.3 Suggested Synthesis

As no software-specific requirements are made by ISO 13485 the synthesis here is simple: follow the requirements made by IEC 62304 as outlined above. Refer to our previous book for further discussion of risk management in general (Juuso 2022).

7.2 Risk-Control Measures [B] [C]

This is where you identified certain software-specific risks involved in the use of your device and you must now consider how to control those risks. You might, for example, decide that the Italian cousin should get an English-speaking sidekick before he goes out to make his first collection rounds around the neighborhood.

7.2.1 Expectations from ISO 13485

In general, ISO 13485 is big on applying the risk-based approach and ensuring that the amount of attention and effort is appropriate to the level of the risk. Thus, it may be seen as fundamentally compatible with the use of the software safety classification used by IEC 62304. No specific requirements for software are made here.

7.2.2 Expectations from IEC 62304

For B- and C-class software, the manufacturer must document risk-control measures (mitigations) for each identified risk where a software item could contribute to a hazardous situation (Clause 7.2.1). This must be in accordance with ISO 14971. Here a non-binding note reiterates that the risk-control measures may be implemented in hardware, software, the working environment, or user instruction.

Clause 7.2.2 specifically addresses risk-control measures implemented as part of the functions of a software item, where it makes the following three requirements beyond what is required by ISO 14971:

- The risk-control measure must be included in the software requirements
- Each software item contributing to the implementation of the risk-control measure must receive a software safety classification based on the very risk the measure is controlling

- The software item must be developed in accordance with the requirements of Clause 5

7.2.3 Suggested Synthesis

As no software-specific requirements are made by ISO 13485 the synthesis here is simple: follow the requirements made by IEC 62304 as outlined above. Refer to our previous book for further discussion of risk management in general (Juuso 2022).

7.3 Verification of Risk-Control Measures [B][C]

After coming up with the perfect mitigation approach, you are also expected to ensure that the plan is acted on. Remember that verification may represent a range of activities from a visual check to an elaborate testing protocol. The risk itself will guide you through addressing each risk control appropriately. In the previous example, Tony might, for example, call up the sidekick to hear whether there have been any issues. If he doesn't like what he hears he might even do spot checks or decide to talk with everyone his cousin has recently interacted with.

7.3.1 Expectations from ISO 13485

No specific requirements for software are made by ISO 13485.

7.3.2 Expectations from IEC 62304

For B- and C-class software, Clause 7.3.1 then considers the verification of the risk-control measures and requires that each identified risk-control measure is verified in a documented manner. The manufacturer is also required to review the measure and determine if it could lead to a new hazardous situation. Also remember that Clause 4.3a instructed the manufacturer to evaluate the effectiveness of any risk-control measure used to lower the software safety classification from B or C (see Section 4.3).

Clause 7.3.2 has been deleted from the amended version of IEC 62304 to reflect the deletion of 7.1.5 above and thus the removed requirement to document sequences of events here.

Clause 7.3.3 addresses traceability via documentation of software hazards, hazardous situations, software items, risk-control measures, and the verification of the measures. Here the manufacturer is required to, as appropriate, document traceability between a) the hazardous situation and the software item, b) the software item and the specific software cause, c) the software cause and the risk control measure, and d) the risk-control measure and its verification. A somewhat cryptic non-binding note offers the ISO 14971 instruction on the risk-management report (its Clause 9 in the 2019 version) as further reading and presumably a good place to ensure traceability has been maintained adequately.

7.3.3 Suggested Synthesis

As no software-specific requirements are made by ISO 13485 the synthesis here is simple: follow the requirements made by IEC 62304 as outlined above. In general, though, ISO 13485 is big on a risk-based approach and ensuring the outcome of actions is evaluated in a way commensurate with the risks posed to the medical device. The link between the software safety classification and the level of rigor applied to verifying risk controls is thus justifiable from the point of view of both standards. Refer to our previous book for further discussion of risk management in general (Juuso 2022).

7.4 Risk Management of Software Changes

> *Once you've built it, don't batch it to smithereens. Both standards realize that a change may not be the improvement it is intended to be, and thus want you to pay adequate attention ahead of making that change. If, for example, Tony decided to switch his cousin for another henchman he thought might do a better job, he should consider the risks involved in the personnel change from the first person going ballistic on hearing he's fired to the replacement then making all the same mistakes.*

7.4.1 Expectations from ISO 13485

ISO 13485 does not make any software-specific requirements here, but it does speak of changes made to the product in general. See Section 8.2 for in-depth coverage of these requirements.

7.4.2 Expectations from IEC 62304

For all classes of software, Clause 7.4.1 instructs the manufacturer to analyze whether a change in the medical-device software (including any SOUP components) a) introduces additional potential software causes contributing to hazardous situations, and b) requires additional software risk-control measures. Note that the latter speaks of software risk-control measures, but it presumably means measures used for the software to control risks, not measures implemented in software as also other forms of measures may be called for.

The clause then makes two additional requirements for class B and C software. Clause 7.4.2 raises the stakes for what was instructed above by Clause 7.4.1a and requires the manufacturer to also analyze whether the changes (to the software or its SOUPs) could interfere with existing risk-control measures. Note that the language also changes from something going wrong (does) to something possibly going wrong (could), and thus the scope is also broader here. Finally, Clause 7.4.3 requires that the manufacturer performs relevant risk-management activities defined in Clauses 7.1–7.3 above based on these analyses.

In other words, the whole of Clause 7.4 compels you to consider any changes made in your medical-device software in terms of the effect it will have on the safety of that device. For all classes of software, this will include an analysis of whether the change introduces new causes for software items contributing to hazardous situations, and whether new risk-control measures are needed. For classes B and C this must also include an analysis of how existing risk-control measures could be affected, and the analysis itself should feed into other risk-management activities described in Clause 7.

7.4.3 Suggested Synthesis

As no software-specific requirements are made by ISO 13485 the synthesis here is simple: follow the requirements made by IEC 62304 as outlined above. Refer to our previous book for further discussion of risk management in general (Juuso 2022).

7.5 The Parts Left Out by IEC 62304

As IEC 62304 in fact builds on the ISO 14971 standard for risk management, and it is to be used on top of the foundations laid by ISO 13485 quality management, there is not a lot the combination of these would not cover. Here IEC 62304 is, however, quite silent on two important aspects: the types of risks you are expected to address, and risk management during the post-market phase of your product.

Addressing the first topic, the types of risks you should consider in the risk management of your device go beyond just malfunctions of the software. IEC 62304 Clause 4.3a alludes to risks beyond malfunctions when it discusses software safety classification and the impact of software – not just software failures – contributing to an undesired outcome, but it is not too vocal on the matter. In practice, you will want to consider at least the following:

- Risks associated with software failures
- Risks associated with the correct, intended use of software
- Risks associated with foreseeable misuse of the software

Note that only the first of these is readily addressed via techniques such as failure modes and effects analysis (FMEA), but all three are to be considered in conducting risk management for medical-device software. The requirement to do so is not made explicit in the IEC 62304, but conformity assessment of the final medical device will raise questions regarding all three types of risks.

On the second topic, IEC 62304 is happy to refer to both production and post-production information in passing. This despite the fact that the IEC 62304 prides itself on being the software life cycle standard. Luckily, ISO 13485 is a little more articulated on the continual nature of risk management. ISO 13485 wants to see risk management as the systematic application of management policies, procedures, and practices to the tasks of analyzing, evaluating, controlling, and monitoring risks (Clause 3.18). It thus opens the lid on post-market surveillance of the device where the IEC 62304 is mostly content with documenting the path to a ready software device and controlling the changes to it later.

IEC 62304, too, acknowledges that software changes may be an occasion to reassess risk management, but it does not go into great detail on what the relevant set of information to consider here might be. ISO 13485, and ISO 14971 in particular, force you to keep your radar on and identify causes for those changes. Similarly, the regulations out there such as the EU MDR (2017/745) will insist you keep an eye on your creation also after it ships.

Note here that ISO 14971 also expects you to verify that any risk controls are effective.

The adoption of the ISO 14971 standard is, for now, an obligatory part of adopting IEC 62304. You must fashion your process according to that standard and also meet the expectations of ISO 13485 but do so while remaining appreciative of the interpretations IEC 62304 wants to see in place.

Chapter 8

CONFIGURATION MANAGEMENT

In some soap operas, the star of the show may change to another actor mid-scene without much fanfare. In software development, this would be considered a case of bait-and-switch. To prevent you from accidentally using one set of deliverables to pass checks and then delivering another set to the next stage, perhaps after some haphazard changes to those deliverables to address failed checks, you are required to pay attention to configuration management.

Configuration management is intuitively understood by people in the industry, but many people will take a pause if you ask them to comprehensively define what exactly a configuration is, and what parts in – and perhaps around – your software should be considered in scope for configuration-management activities.

The ancestral standard, ISO 9000, takes the responsibility of defining what a configuration is in our context. In Clause 3.10.6 of the standard, the term is defined as the interrelated functional and physical characteristics of a product (or service) defined in the product configuration information. The configuration information itself may be in the form of requirements or other information (i.e., documentation of some sort) for product design, realization, verification, operation, and support. Thus, the configuration is to be understood as the grand constellation of your software – the planets, stars, dust, and dark matter (i.e., SOUPs) that make up the whole of your software. A distant cousin, the *IEEE Standard Glossary of Software Engineering*

Terminology, takes perhaps the most concrete view of a configuration item by calling it an aggregation of hardware, software, or both that is treated as a single entity in your scheme. In some contexts, you might thus even call the Milky Way a configuration item if it made sense in your grand scheme of things, but you would not expect it to just be one item if you dove deep down into it. Imposing any configuration management over a celestial body such as the Milky Way which moves at a velocity of approximately 1.3 million miles per hour is probably a topic best left to astronomers, though.

ISO 13485 is oblivious to the whole concept of a configuration but accepts the ISO 9000 definition at face value through a high-level normative reference made to the standard. IEC 62304 appears to take a live-and-let live approach to this definition (i.e., no reference to it is made whatsoever) and adopts somewhat of a politician's approach to then defining the term "configuration item" as some entity uniquely identifiable at a given reference point (Clause 3.5). Your whole software must be controlled as an entity, and a version change on that level will require software configuration management action (note to Clause 3.34). The standard also requires that all configuration items for B- and C-class software items are placed under configuration control prior to verification activities – thus configuration management is required to step in by the verification stage or just before release for class A software.

The aim of configuration management is to understand what elements you need to control, when, and how. The point of the exercise is to both identify the constituent parts and to manage changes to the whole. In addition, the environments, platforms, tools, and equipment that touch your software product may also be a part of your considerations here.

Both standards want to see changes being controlled by appropriate reviews taking place before implementation and the corresponding records being kept afterward. IEC 62304 is again more articulated in terms of what this means in the context of software, but ISO 13485 does not sit silently either. Now ISO 13485 has requirements on what it wants to see take place in the context of changes being made, and it not only wants to see this evidence for software but also for any other type of medical-device product.

The topics covered in this clause are given in Table 8.1 below. The table also notes which software safety classes each topic is applicable to, although in this case, all the requirements apply to all software safety classes. Note, though, that the point during your software-development life cycle after which configuration management must apply may vary based on your software safety classification (see Section 8.1.2).

Table 8.1 Topics of Configuration Management

#	Topic	A	B	C
8.1	Identification of configuration items	☑	☑	☑
8.2	Change control (incl. verification)	☑	☑	☑
8.3	History of controlled items	☑	☑	☑
8.4	The parts left out by IEC 62304	☑	☑	☑

☑ = all apply, ☒ = some apply, ☐ = none apply

The clause has an immediate overlap with ISO 13485 Clause 7.3.9 on control of design and development changes. In addition, some foundational requirements related to configuration management are made by the standard's Clauses 7.5.8 on identification and 7.5.9 on traceability. These requirements will be discussed next.

8.1 Identification of Configuration Items

Looking back, Michael J. Fox may be the only choice for the lead in the Back to the Future *trilogy, but he was not the first actor hired to play the role. For a time, Sylvester Stallone, too, was set to walk in Eddie Murphy's star-sprinkled tennis shoes in* Beverly Hills Cop *– hence the Stallone poster behind Rosewood's door in the film. These are not the only replacement players destined to walk the unlikely paths they ended up taking. These examples, however, show that casting matters, and in software development, this is known as the management of the software-system configuration.*

Identification of configuration items has, in effect, two relevant points of view. IEC 62304 primarily looks at the version-controlled items that go into your product, how these are identified, and what processes may be used to enact any changes in them. The standard even goes as far as to define the term "version" as an identified instance of a configuration item (Clause 3.34), thus tying a knot between the two concepts even if the term version has broad use and is understandable based on even a dictionary definition. A nonbinding note pulls the knot tighter by asking for the software configuration process to be activated when modifying a version of medical-device software.

ISO 13485 also appreciates the changes going into your product, in much the same way as IEC 62304, but it perhaps steps back a little to address the traceability of all your records and activities and the finished products you ship. Both standards do want you to stay on top of both changes in the software and the changed software going out. The former entails the use of change requests and the latter some agreement with you, your customers, your notified body, and your other regulatory authorities on what level of changes require what types of actions.

The relevant clauses of both standards are:

- ISO 13485 Clause 7.3.9 Control of design and development changes
- ISO 13485 Clause 7.5.8 Identification
- ISO 13485 Clause 7.5.9 Traceability
- IEC 62304 Clause 5.1.9 Software configuration management planning
- IEC 62304 Clause 8 Software configuration management process

In the following, we will look at the expectations of both standards, one standard at a time, and then synthesize a plan to meet the requirements of both.

8.1.1 Expectations from ISO 13485

As discussed above, the ISO 13485 standard does not speak of configurations specifically although it does expect all D&D changes to be identified (Clause 7.3.9). The general requirements it makes on maintaining the necessary identification (Clause 7.5.8) and traceability (Clause 7.5.9) information are defined on a high level in the standard. The requirements in these clauses cut through all your operations, and thus apply equally to the documentation of your activities and to any medical devices placed on the market, including any changes made to them. These themes are discussed in detail in our ISO 13485 book (Juuso 2022).

8.1.2 Expectations from IEC 62304

IEC 62304 devotes the whole of its Clause 8 to the topic of configuration management, and it also addresses the requirements for software configuration-management planning earlier in Clause 5.1.9 (see Section 5.1). Clause 8.1.1 builds on this platform and requires you to establish a unique identification scheme for configuration items and their versions. This scheme must cover other software products, SOUP components, and documentation, and it is to match with the development and configuration planning required by Clause 5.1. This requirement applies to all software safety classes.

In terms of SOUP components (incl. standard libraries), you must document at least the title and the manufacturer of the component (Clause 8.1.2a–b). For uniquely identifying the component you may, for example, use the software version, release date, patch number, or upgrade designation.

In terms of documentation, you are required to identify and document the configuration items (incl. their versions) that comprise the software system configuration (Clause 8.1.3). Also note that Clause 5.1.11 requires that for class B and C software, all configuration items must be placed under configuration control before they are verified. Finally, remember that Clause 6.2.5 (see Section 6.2) also requires you to communicate with your users and regulatory authorities on problems and modifications.

8.1.3 Suggested Synthesis

The requirements of the two standards are complementary through and through. Table 8.2 provides a glance over the main requirements of both standards. Note the software safety classification used by IEC 62304 and remember that some clauses of ISO 13485 may be excluded with a valid rationale (see Juuso 2022).

The fundamental expectation by both standards is that you have a process for configuration management. To make that process more practical, we would suggest considering the following regarding how to implement version numbering and what items to manage via configuration management.

Table 8.2 Synthesis of Top-Level Requirements

Combined requirements	ISO 13485	IEC 62304		
		A	B	C
A process for identifying product and product components	✓	✓	✓	✓
Identify and document the configuration items (incl. versions) that comprise the software-system configuration	-	✓	✓	✓
Place software under configuration control prior to verification	-	-	✓	✓

8.1.3.1 Version Numbering

The first question to solve in designing your overall approach to configurations is how you intend to identify the versions of the individual software modules that go into your software system, and how you designate the versions of the final software being released.

For the finished product, we would advise the use of a four-level version numbering scheme that will allow you to address the magnitude of the changes you are contemplating in terms easily understood by regulations. This model might, for example, use a four-digit scheme (e.g., A.B.C.D.) to observe the following triggers.

- **A-level change**

    Any significant change to the intended use or indications of your product. This will be a point of serious introspection where you consider, among other relevant aspects, the validation, risks, and benefits associated with your software.

- **B-level change**

    A significant change to the risk management of your product. This might, for example, involve new risk-control measures (mitigations) or changes made to the software that affect the risks and benefits associated with it.

- **C-level change**

    A feature update to the software within the confines of the present intended use, indications, and risk management.

- **D-level change**

    A security patch or a bug fix to the software.

The above scheme will speak directly to the regulatory requirements for your product and will help you in placing the needed emphasis on the major changes (A and B) while attending to more minor changes (C and D) more efficiently. You may be able to use this granularity in conversing with your regulatory body, e.g., on what level of changes need to be reported ahead of time, and what changes may require new conformity assessments (e.g., for IEC 62304 conformance).

Alternatively, you might, for example, employ a three-digit scheme based on the widely popular semantic versioning scheme that is coupled with changes on the level of major updates (A), minor updates (B), and bug fixes (C). You might even use one scheme on the level of your finished product and another on the level of the software items comprising that product. This way, a change on a lower level does not as directly translate to a change of the same magnitude on the level of the final product. You must evaluate the impact of the change in terms of the final device, but a major change in some low-level component does not always indicate a major change on the level of the integrated whole. Occasionally it may also be that a minor change on a lower level leads to a major change on the top level depending on the use of that component. For this reason, it may make sense to decouple versioning on the product level and the component level. Also remember that for any SOUP components you are, of course, more or less stuck with the versioning scheme their manufacturer has chosen.

Neither of the above schemes, the three-digit or the four-digit scheme, is required by either standard. You must have a way of identifying versions, but how you do so is up to you. You may set up your versioning scheme as you see fit. Remember also that you may later want to develop new versions of the sold software both as descendants of any current version, but also as forked versions to meet any number of emergent needs (e.g., bug fixes, feature tweaks, cheaper/more premium versions of the same software). Ideally, your configuration management should support all these goals. Finally, if possible, you should also ensure that you are able to differentiate between changes that have not gone outside your door yet and those which have been supplied to customers. This may be a register matter, but it may also involve addon monikers such as "RC" for a release candidate.

8.1.3.2 What Items Should Be Tracked?

IEC 62304 Clause 8.1.1 requires you to establish a scheme for identifying and controlling configuration items (incl. their versions), but it stops short of specifying a closed list of what these items are. The objective here is not only to know what the constituent parts of the medical-device software are, but also to control any changes in them. In theory, Clause 8.1.1 leaves the door open for the manufacturer to decide what is and what is not a configuration item but arguing that something doesn't need to be controlled may be an uphill battle. In practice, you will want to control everything, but the level of control you impose on any given type of item will depend on its risk profile.

From IEC 62304, it is quite clear that the software itself and any software items it consists of, that is relevant objects of source code, should be identified on some meaningful and practical level. This also applies to entities such as SOUP components and documentation.

In terms of documentation, both standards want to see control exercised over the documents and records you have. Changes to the documentation may be triggered by a change in the corresponding software or by some identified need for a change in the documentation itself. Any changes to the code must, of course, be weighed for their impact and any appropriate action, including updating the corresponding documentation, must be taken. If, on the other hand, there is a separate need to only change the accompanying documentation, e.g., by fixing a small typo in it or by improving upon the language somehow this should be managed via an appropriate review and approval mechanism and versions maintained. You will want to record the change using your configuration-management process, but it would be difficult to argue why some documentation has evolved in a completely different way than it was originally created. Also, remember that the baseline for revising documentation is given by your document and record control procedures as instructed in your QMS (see Juuso 2022). In the case the documentation in question is part of the product, IEC 62304 wants to see changes recorded via your configuration-management processes. A note to Clause 5.1.8 also points out that the whole of Clause 8 is relevant when considering configuration management of documentation.

Beyond code and documentation, the line becomes more subjective. The IEEE 1042 guide to software configuration management, for example, offers up a much more detailed list than what is part of either of the standards. Here the entities to be considered cover, for example, the following:

- Documentation, including management plans, specifications (e.g., requirements and design), user documentation, test designs (incl. specified test data), and maintenance documentation
- Code, including source code, executable code (i.e., binaries), libraries
- Support software
- Data dictionaries, various cross-references, and databases (both included and processed data)

Much of the above list is addressed by ISO 13485 quality management and IEC 62304 software development (e.g., plans, documentation, code, libraries, and support software), but notably data dictionaries, cross-references, and databases may not be an instantly apparent part of the same picture. Test data, too, may fall under this latter group, but we feel that test data is managed by testing activities and their reviews, and thus is not a part that ships with the configuration in any active form, although it too most certainly needs to be controlled. In the following, we address this group of three entities together as databases.

8.1.3.3 Configuration Management over Databases

One particularly complex class of components to consider here is the databases used by your software. Depending on the nature of the databases, they may have as much impact on your device as the code itself. The databases may also evolve over time, and thus change the functioning of your device over time. The impact of data and databases may be even more pronounced if the device somehow uses machine learning or artificial intelligence methods based on the data.

Databases may be a part of the medical device, or they may be placed outside the medical device and used via some interface. The placement of the database in terms of what is in and outside of the device is a critical design choice here. Think of the architectural design of your software, but also think about the intended use of your device and what the role of the database is in achieving that use when deciding whether the database is in or out. Of course, something placed outside of the device is not something that could be regarded as out of sight, out of mind. Based on the use of the database and its impact on the risk assessment of the device you will want to address it differently.

Databases contained within the product should be controlled as configuration items in a way commensurate with their risk profile: a database of log entries may, for example, carry smaller risks than some databases used for machine learning or deep learning activities and affect the performance of the device directly. In the former case, the database may perhaps be considered on the level of its schema and allowed to be populated via the normal use of the device. In the latter case a much more involved re-evaluation, perhaps even including an updated clinical evaluation, may be called for. In between the use of databases may be akin to the use of SOUP components.

Databases used via interfaces and not part of the device itself should be acknowledged in the description of the device (perhaps even in the intended use), and all care should be taken to assess and monitor their use, but they are not a part of the configuration management of the device itself in the same way as a constituent software component. In this case, the use of the database may be part of the requirements analysis, where the requirements for using the interface (incl. identity, data formats, protocols, and quality of service parameters) are addressed carefully.

In figuring out whether a database, or indeed some other item, should be brought into the domain of configuration management ask yourself the following.

- Can a change in the functioning of the database affect the conformity of the device with its requirements (incl. applicable regulations)?
- Is the functionality enabled by the database considered to be medical in nature? In the EU, see the MDCG 2019-11 guidance, for example.
- Is the database related to achieving the intended use of the device?
- Is the database included inside the device in the architectural design?
- Is the database version controlled so that a version 1.0 and, for example, a version 2.0 can be uniquely identified at a given reference point?
- Is the database used to modify the functionality of the device? For example, via machine learning and artificial intelligence methods.

If your answer to any of the above is yes, you will more than likely be looking at a database that should be brought into configuration management. If your database is implicated in meeting the intended use of the device, is architecturally located within the device, and can be uniquely identified in terms of its version, you will want to address it via configuration management. The last condition will no doubt sound sloppy, but it is how IEC 62304 defines a configuration item.

Regardless of whether the database is a part of your device, many issues may potentially be caused by unexpected changes in a database, and these may need to be reflected in your requirements analysis and risk management. Examples here include a change in the format of a database field (e.g., integer versus float), a change in the number or order of database columns (e.g., items referred to by ID versus items referred to by sequence), a change in the expected protocol of an interface, or a silent change in how a database field is used. Particularly the last example may be a troublesome source of errors because of a silent drift in how people enter field values and some ambiguity exists in what exactly is expected. The same may also be true for a database fed by web-connected sensors or other IoT-devices: a new brand or model of a device may have different characteristics from those used before and may thus provide measurements that are not exactly in line with the measurements obtained previously. Again, the above issues may be particularly problematic if the retrieved data is used to guide the operation of the software, for example, via machine learning methods.

8.2 Change Control (Incl. Verification)

Hollywood types call their agents and lawyers when a change in casting is needed, but for software, change control starts with a softer concept: a change request. Here the objective is to ensure that only evaluated changes are enacted and that the necessary records are kept.

The previous clause went into detail on how to set up the GPS system for identifying the parts, and the whole, of your software. If your software was built once and never updated in any way that coordinate system would not be needed for much. As your software does change, that system will, however, get its chance to affect the marquee signs. That is why the second part of Clause 8 goes into detail on how you manage changes in your identified configuration items.

The relevant clauses of both standards are:

- ISO 13485 Clause. 7.2.2 Review of requirements related to product
- ISO 13485 Clause 7.3.9 Control of design and development changes
- ISO 13485 Clause 7.3.10 Design and development files
- IEC 62304 Clause 8.2 Change control

In the following, we will look at the expectations of both standards, one standard at a time, and then synthesize a plan to meet the requirements of both.

8.2.1 Expectations from ISO 13485

The fundamental expectation from Clause 7.3.9 is that you have a documented process for controlling D&D changes. The trigger for the activity here may be a minute change to the inner workings of a software module but the point of study is placed squarely on how that affects the overall product. Here you should apply your understanding of the regulatory landscape around your product and the expectations of your regulatory authorities to arrive at a flowchart of what types of changes call for what types of activities.

The significance of the proposed change to the function, performance, usability, safety, and regulatory requirements applicable to the medical device and its intended use are the prime concerns here. The changes must be reviewed, verified (see the section below on how IEC 62304 understands verification in this context), validated (as appropriate), and approved before implementation. The review must also include an evaluation of effects on constituent parts, products in process, or already delivered, and the inputs or outputs of risk management and product realization. Records of the changes and their review (incl. any necessary actions) must be kept.

Also, note that Clause 7.2.2 requires you to confirm customer requirements before acceptance if no documented requirements are provided by the customer, and you must also ensure that when product requirements are changed the relevant personnel are made aware. Remember that changing requirements may also have a trickle-down effect throughout your software-development activities, such as the revision of your risk management file.

Finally, Clause 7.3.10 expects that a D&D file is kept for each medical-device type (or medical-device family) and that this includes or references records generated to demonstrate conformity to D&D requirements and any D&D changes. This short clause thus expects D&D changes to be on record.

8.2.2 Expectations from IEC 62304

8.2.2.1 Approval of Change Requests

Changes to the identified configuration items (see Section 8.1) may only take place in response to approved change requests (Clause 8.2.1). For this reason, it is impractical to drag out the approval of a change request form in the hopes of recording absolutely everything from the implementation of that change with the same form. This is occasionally attempted by organizations, but it flies in the face of the expectations of the standard. Don't do it.

Note that the approval process may be defined as part of your change-control process or elsewhere in your QMS processes. Different approval processes may also be used, for example, depending on the stage of the development (see Clauses 5.1.1d and 6.1e in Sections 5.1 and 6.1, respectively). All that is fundamentally required here is that no changes are implemented prior to approval and that you have some process for performing the appropriate approval.

8.2.2.2 Implementation of Changes

The implementation of the approved change is to take place as prescribed in the change request (Clause 8.2.2). You are required to identify and perform any activities that need to be repeated due to the change, including any changes to the software safety classifications on any level (from software systems to software units, all of which are software items). A note to the clause emphasizes that the actual implementation process does not need to be covered by the change-control process but can instead refer to your other processes (see Clause 5.1.1e and 6.1e). There is no need to reinvent the wheel here, and little point in maintaining multiple parallel implementation processes.

8.2.2.3 Verification of Changes

Here the standard uses the term verification quite loosely: what is required is that the changes are verified, not that some major verification activity is necessarily performed. You are required to verify the change, and if some earlier verification activity is rendered invalid by the change, redo that verification while taking Clauses 5.7.3 and 9.7 into account (Clause 8.2.3). What is meant by verification here is thus not the same as your overall verification activities unless the change made calls those activities into question somehow.

8.2.2.4 Traceability over Changes

Finally, Clause 8.2.4 requires you to maintain traceable records of the relationships and dependencies between change requests (incl. any relevant problem reports) and their approval. These are also to be available somehow for later accounting as discussed in Section 8.3.

8.2.3 Suggested Synthesis

The requirements of the two standards are complementary through and through. Table 8.3 provides a glance over the main requirements of both standards. Note the software safety classification used by IEC 62304 and remember that some clauses of ISO 13485 may be excluded with a valid rationale (see Juuso 2022).

The fundamental expectation by both standards is that you have a process for change control. IEC 62304 requires that changes to the software product are only made based on documented specifications in the form of change requests. ISO 13485 on the other hand requires you to have a process for controlling all design and development changes. The result is

Table 8.3 Synthesis of Top-Level Requirements

Combined requirements	ISO 13485	IEC 62304		
		A	B	C
A process for controlling D&D changes	✓	✓	✓	✓
Control changes impacting the function, performance, usability, safety, and the regulatory requirements applicable to the medical device and its intended use	✓	-	-	-
Control changes over other software products, SOUP components, and documentation	-	✓	✓	✓

that you need to have a process using change requests in place, and the place to define that process is in your SOPs (e.g., in SOP-11 as per Juuso 2022). The process should then ensure that any changes are assessed before being enacted, verified after enacted and that appropriate records are maintained.

8.3 History of Controlled Items

> *Sylvester Stallone does not list* Beverly Hills Cop *in his filmography, but if the film was a medical device, he might have to live with the fact that somewhere in its documentation he would be listed as one of the actors that went through that particular version of police academy. Similarly, with medical software, you are required to keep track of its pedigree.*

Here ISO 13485 is yawning and insisting that we already covered this ground when setting up the whole of your quality management system. IEC 62304, on the other hand, thinks you may suffer from dementia and hammers home the need to maintain traceable records.

The relevant clauses of both standards are:

- ISO 13485 cl. 7.3.9 Control of design and development changes
- IEC 62304 cl. 8.3 Configuration status accounting

In the following, we will look at the expectations of both standards, one standard at a time, and then synthesize a plan to meet the requirements of both.

8.3.1 Expectations from ISO 13485

As previously covered in Section 8.2, ISO 13485 expects records to be kept of the changes and their review (incl. any necessary actions). In addition, the baseline for controlling documentation is given by your document and record control procedures as instructed in your QMS (see Juuso 2022).

8.3.2 Expectations from IEC 62304

Clause 8.3 requires that you retain retrievable records of the history of controlled configuration items (incl. system configuration). This requirement applies to all software safety classes. Clause 8.3 is called Configuration Status Accounting, which almost implies a need to audit the status of the configuration occasionally, but what is actually meant here is that the records enabling such an audit are maintained. No status review is required here, only that you retain the appropriate records.

Expect to retain all the records you specified above in Sections 8.1 and 8.2, but note that no list is in fact given here. Clause 8.2.4 discussed above did, however, name the change request, problem report, and approval of the change request as records to maintain and link together.

8.3.3 Suggested Synthesis

The requirements of the two standards are complementary through and through. Table 8.4 provides a glance over the main requirements of both standards. Note the software safety classification used by IEC 62304 and remember that some clauses of ISO 13485 may be excluded with a valid rationale (see Juuso 2022).

Here the impact on your processes and records is nothing new: you already addressed it all above.

Table 8.4 Synthesis of Top-Level Requirements

Combined requirements	ISO 13485	IEC 62304		
		A	B	C
Maintain traceability of the records	✓	✓	✓	✓

8.4 The Parts Left Out by IEC 62304

Could it be that there's really nothing more to add here? The above discussion already covered everything expected by the two standards in an exhaustive manner. Perhaps the only comment to add here is that configuration management is not an afterthought to meeting the standards, and it is not something you should only see to once the cracks caused by rust and icebergs alike start to affect the maneuvering of your ship. Keeping track of those patches and dented parts of the hull is required all along. It is, however, a fairly easy task to get done correctly: it is something which if done correctly allows you to track your deliverables and the work done on them efficiently, securely, and predictably. If your system works for you and answers all of the needs discussed above, then you should be well-set.

Change control is something then built on top of configuration management, to facilitate the change undertaken within that constellation. The first objective here is naturally, as discussed above, to ensure that a seaworthy ship stays seaworthy throughout any later changes you make. The change documentation will then be viewed in any audits of the device to assess how it has evolved and what types of pressures have been inflicted upon it. The flip side of this is that your processes for change control are the very mechanism that makes it possible to quickly evolve your device in response to any identified issues, demands, and appealing new opportunities. Setting up a lean, efficient, and safe process for making changes will be to your advantage from both a business and a regulatory point of view. It will also generate a significant part of the information you will exchange with your regulatory authorities over the coming years, and as such is already the object of much attention from the point of view of your regulators. Making your change control processes unnecessarily burdensome at your organization will prevent you from carrying out some changes and will mean you misapply your attention during making the changes you do make – it may also mean that your regulatory authorities drown under all of the noise generated by the manufacturers under their control. Ensuring change control achieves the objectives set out above in a lean, efficient, and effective manner is in fact what both sides of the table want to enable safe, ever better medical devices. You can thus expect more guidance, standards, and regulation to come out in this space in the coming years.

Chapter 9

PROBLEM RESOLUTION

Houston, we have a problem. Apollo 13 is perhaps the best example in popular culture about encountering problems, analyzing the situation, coming up with a workable solution, and saving all the human lives at risk. The Titanic hitting that iceberg may be another example, but one with less success in any problem-resolution process. Problem resolution is about what you do when more than hot air hits that fan. It's also that last failsafe lifejacket your auditors look to when assessing your otherwise perfect operations – or the preplanned startup sequence that ensures your spaceship can power back up for Earth re-entry.

Problems will occur. What you then do makes the difference between brushing the issues under the carpet and keeping them from becoming epidemics. It's easy to think that a perfect system would make zero mistakes, especially as you are starting out and especially if you are looking to focus on the code itself and do as little of everything else around it as possible. Not documenting issues is an issue. Not fixing issues is an issue. Not fixing issues quickly, though, is less of an issue than not fixing issues correctly. In recent years there have been some high-profile cases where regulatory bodies have been moved to jolt even large manufacturers into faster action as their recalls have slowed to a crawl despite having vast resources to deal with the issues. With big resources usually comes big business and large user bases, and as a result, even small problems may become behemoths in this geographic space. For this reason, if you are a small manufacturer your resources may be better for dealing with a problem than if you are a large manufacturer, even if that seems counterintuitive at first.

But what is a problem? Neither standard defines the term, so the standard dictionary definitions apply here. IEC 62304 does define the term problem report as a record of actual or potential behavior of medical-device software that a user or some other interested party believes to be unsafe, inappropriate for the intended use, or contrary to the specification (Clause 3.13). Note that the definition of software here includes programs, procedures, and associated documentation and data. Thus, a problem may be seen as a mismatch between what you got as a product and what was expected based on the intended use and specification. The standard then makes nonbinding notes to explain that a problem can concern a released product or a product still in development and that in the former case, Clause 6 makes additional requirements on the handling of the problem regarding regulatory actions (see Section 6.2).

Your problem reports, then, should not start with "Once upon a time…" unless you are writing a fairytale that you don't expect anyone to actually believe. Your problem-resolution process should not be about finding a patsy to blame, nor should it always end with a statement that this was a one-time occurrence that was a statistical anomaly in your otherwise perfect software. Your problem report should definitely not blame the user as a go-to answer, and it should not say that you never thought of the software as finished in the first place. Others have tried all these excuses, but as excuses go, they are pitiful. The first step to fixing a problem is admitting you have one, right?

This section will take you through the eight steps the IEC 62304 standard sees as needed to resolve problems appropriately. The section also contrasts these with the expectations of the ISO 13485 standard, as in earlier sections, and provides a synthesis meeting the requirements from both. The topics covered in this clause are given in Table 9.1 below. The table also notes which software safety classes each topic is applicable to, although in this instance all the requirements apply to all classes of software.

The requirements from IEC 62304 are thus set in a convenient chronological order, whereas the related requirements from ISO 13485 are to be found in a few different clauses of the standard. Let's look at the steps involved in problem resolution one at a time.

9.1 Prepare Problem Reports

After admitting you may have a problem, the first step is to record that problem so that it can be analyzed, corrected, and all of the

Table 9.1 Topics of Problem Resolution

#	Topic	A	B	C
9.1	Prepare problem reports	☑	☑	☑
9.2	Investigate the problem	☑	☑	☑
9.3	Advise relevant parties	☑	☑	☑
9.4	Use change-control process	☑	☑	☑
9.5	Maintain records	☑	☑	☑
9.6	Analyze problems for trends	☑	☑	☑
9.7	Verify software problem resolution	☑	☑	☑
9.8	Test documentation contents	☑	☑	☑
9.9	The parts left out by IEC 62304	☑	☑	☑

☑ = all apply, ☒ = some apply, ☐ = none apply

related parties can hear about the happy conclusion. This is where you play Columbo and attempt to record all the facts after walking into the room with the chalk outline on the floor.

Clause 9.1 in IEC 62304 takes a cautiously pragmatic approach to problem resolution and instructs how the problem is to be recorded. First and foremost, it wants to see that any problems are in fact recorded by the manufacturer, but it does list some information it wants to see included in the report. ISO 13485 chimes in to the extent that we may be talking about nonconforming products or processes.

The relevant clauses of both standards are:

- ISO 13485 Clause 8.5 Improvement (incl. CAPA)
- ISO 13485 Clause 8.3 Control of nonconforming product
- ISO 13485 Clause 8.2.2. Complaint handling
- IEC 62304 Clause 9.1 Prepare problem reports

Also note that Clause 7.3.10 on D&D files instructs records on conformity to D&D requirements and D&D changes to be maintained, which is something you may be acting on here either by investigating problems or making changes. No specific requirements are made here on the kind of information you will record. You could also find other related sections in ISO 13485, for example, Clause 8.2.2. on complaint handling may describe a similar process of recording a report, investigating it, and proceeding to fix any issues. If comparing and contrasting with these processes somehow simplifies your different approaches here go ahead but know that this is not required. Not all customer feedback is on problems regarding your software. Not every software problem is identified through customer feedback. We would suggest that not every complaint represents a software problem needing to be fixed, but those that do may then lead to even multiple CAPA actions. For this reason, these clauses are listed here but covered only briefly below. You will find additional discussion on the clauses in our previous book (Juuso 2022). In the following, we will look at the central expectations of both standards, one standard at a time, and then synthesize a plan to meet the requirements of both.

9.1.1 Expectations from ISO 13485

In general, ISO 13485 wants to see any encountered problems weeded out, whether these are detected in the products or the processes of the QMS. It does not acknowledge a problem report per se, but it does speak profusely about nonconformities (Clause 8.5) and nonconforming products (Clause 8.3). It also goes to great effort to point out that such problems may be detected through several seemingly separate activities, such as customer feedback and complaints (e.g., Clauses 7.2.3c and 8.2.2), lowly servicing information (Clause 7.5.4a), design and development work, and the long-term post-market surveillance activities. The area of most immediate common ground with IEC 62304 in terms of problem resolution is in the control of nonconforming products (Clause 8.3) detected before or after delivery, but equally a complaint report (Clause 8.2.2) could serve as the beginning of a problem report.

The focus in Clause 8.3 is on faulty products that should first be sequestered and separated from sellable products and then investigated to find the cause of the fault and an appropriate remedy if there is one. The clause does not directly translate to a problem report as discussed in IEC 62304, but it is nonetheless worthwhile to know what the standard expects to see regarding problem handling when it involves faulty products. ISO 13485 Clause 8.3.1 further requires that you have a process for nonconforming products (incl. controls, responsibilities, and authorities for their identification, documentation, segregation, evaluation, and disposition). This may factor into the consideration of what information you want to record in your problem report or other subsequent handling documentation. See SOP-2 in our earlier book

(Juuso 2022) for more on handling nonconforming products, or SOP-7 on feedback and complaints.

If the problem represents a nonconformity to be addressed, you should invoke your corrective actions and preventive actions process (CAPA; see SOP-2 in Juuso 2022). If on the other hand, the problem does not meet your definition of a nonconformity, e.g., the problem represents an issue in as-of-yet unreleased software, a leaner process may suffice.

9.1.2 Expectations from IEC 62304

Clause 9.1 requires the manufacturer to prepare a problem report for each problem detected in medical-device software. This must include the following:

- Statement of criticality (e.g., effect on performance, safety, or security)
- Other information that may aid in the resolution (e.g., devices and accessories affected)

Note that prior to the 2015 amendment, the clause also called for the type of the problem report (e.g., corrective, preventive, or adaptive to new environment) and the scope of the problem (e.g., size of change, number of device models/accessories affected, resources involved, or time to change) to be recorded. Now these properties may be thought of as included in the category of other information as needed.

A note to the clause points out that problems can be discovered before or after release, and both inside and outside the manufacturer's organization. This corresponds nicely with the discussion of before and after delivery by ISO 13485, although release and delivery are distinctly different events.

9.1.3 Suggested Synthesis

The requirements of the two standards are complementary through and through. Table 9.2 provides a glance over the main requirements of both standards. In this case, the software safety classification used by IEC 62304 does not really change matters, and you might be hard-pressed to argue that the clauses of ISO 13485 should be excluded (see Juuso 2022).

The requirements here are quite brief, and once again IEC 62304 would be happy with the correct outcome even without a previously defined SOP.

Table 9.2 Synthesis of Top-Level Requirements

Combined requirements	ISO 13485	IEC 62304		
		A	B	C
A process for nonconforming products	✓	-	-	-
A problem report	-	✓	✓	✓
A statement of criticality	-	✓	✓	✓
Other information that may aid in the resolution	-	✓	✓	✓

Since you will have related SOPs from ISO 13485 you will do your best to ensure their instruction meets both sets of requirements. Any instruction in the SOP should address the problem report as discussed above.

9.2 Investigate the Problem

This is where you put your notepad back in the pocket of your crumbled-up trench coat, absentmindedly thank your first interviewees, and go off to massage those grey brain cells to figure out what has happened. Unlike in daytime detective stories, the objective here is not to outsmart the murderer but instead figure out what in your product (or processes) should be fixed and how to avoid any harm coming to your users.

After you have a problem report, it's time to investigate it. If the report is gibberish, as in clearly the result of a misunderstanding, or made in error you should be able to close the case here before proceeding further. If it is a valid report of a problem, you will be expected to look to your processes for risk management, CAPA, nonconforming product, and act appropriately.

The relevant clauses of both standards are:

- ISO 13485 Clause 8.5 Improvement (incl. CAPA)
- ISO 13485 Clause 8.3 Control of nonconforming product
- IEC 62304 Clause 9.2 Investigate the problem

In the following, we will look at the expectations of both standards, one standard at a time, and then synthesize a plan to meet the requirements of both.

9.2.1 Expectations from ISO 13485

As with the previous clause, here too the most directly relevant requirements for investigation of the problem are defined by ISO 13485 Clause 8.3.1, which we also used in the previous section. The requirements added here are as follows:

- You must determine the need for an investigation and notification of any external parties responsible for the nonconformity
- You must record the nature of the nonconformity and any actions taken (incl. investigation and rationale for decisions)

Note that these requirements are very similar to what is said about complaint handling in Clause 8.2.2. Your processing of complaints is required to be timely in accordance with applicable regulations, any resulting CAPAs are to be recorded, and should you decide not to investigate a complaint you must record a reason. The clause also gives you a minimum set of items to record for the complaint: a) receiving/recording information, b) evaluation of the feedback as a complaint, c) investigating complaints, d) determining the need for reporting to regulatory authorities, e) handling of related product, and f) determining the need for CAPA. If the problem arises out of a complaint and represents a nonconformity to be addressed, you should follow your communication processes and invoke your corrective actions and preventive actions process (see SOP-7 and SOP-2 in Juuso 2022).

All of the above may factor into the consideration of what information you want to record in your problem report or other subsequent handling documentation. You do not need to include all this information in a first-line problem report, but you may want to record the information at an appropriate place and time during your problem handling if a released product is involved.

9.2.2 Expectations from IEC 62304

Clause 9.2 compels the manufacturer to a) investigate the problem and, if possible, identify the causes, b) evaluate the relevance to safety using their risk-management process, c) document the outcome of the investigation and evaluation, and d) create change requests for actions needed to correct the problem. In case no action is taken in d) the rationale for doing so must be documented. A nonbinding note to the clause further explains that if the problem is not relevant to safety, a correction is not required for complying with Clause 9. Similarly, a nonbinding note to the definition of the term problem report in Clause 3.13 (see the beginning of Section 9 above) clarifies that a problem report may be rejected if it represents a misunderstanding, error, or insignificant event.

9.2.3 Suggested Synthesis

The requirements of the two standards are complementary through and through. Table 9.3 provides a glance over the main requirements of both standards. In this case, the software safety classification used by IEC 62304 does not really change matters, and you might be hard-pressed to argue that the clauses of ISO 13485 should be excluded (see Juuso 2022).

Your process for investigating problems should most likely lean on your already-defined CAPA process (see Juuso 2022) so as to not create multiple, deceptively similar processes for similar purposes. The general approach to CAPA with its emphasis on identifying the root cause should also work here in the context of software. A quick glance over Table 9.3 confirms that all safety classes of software are in fact treated the same here, and thus you may now ensure that your general ISO 13485 CAPA process also suits your needs in addressing software-related problems. A problem suspected in high-risk software is going to be more worrisome than a similar problem suspected in low-risk software, but in the eyes of the requirements stated here by IEC 62304 there is no difference.

Table 9.3   Synthesis of Top-Level Requirements

Combined requirements	ISO 13485	IEC 62304		
		A	B	C
A statement on the need for an investigation	✓	✓	✓	✓
A statement on the notification of external parties	✓	-	-	-
Identify the causes (if possible)	-	✓	✓	✓
Evaluate relevance to safety using risk management	-	✓	✓	✓
Document the investigation/evaluation outcome	✓	✓	✓	✓
Record of the nature of the nonconformity and actions taken	✓	-	-	-
Create change requests for actions needed	-	✓	✓	✓

Note that ISO 13485 seemingly allows for deliberation on the need for an investigation, but IEC 62304 requires the manufacturer to investigate the problem. The difference here appears to be semantics as at least a record and analysis of the problem is in any case required by both standards. Putting its fight face on, IEC 62304 now requires you to use your risk-management process to investigate the problem's relevance to safety. At a minimum, we might interpret this to mean investigating the problem in light of your risk management file for the device and reacting to any action and revision needs from therein, but in the face of a starker problem, you might even trigger a new ISO 14971 risk-assessment project as per your SOP. You may, of course, find in the course of resolving the problem that you need to also revise your risks or their mitigations somehow, which in turn might call for communication with external parties, as covered next.

9.3 Advise Relevant Parties

All good detective stories end with all the interested parties couped up at some final gathering where the master detective carefully unwinds the story as he has put it together, and finally takes a victory lap around the room after naming the person who did it. The standards don't care about egos and dramatic storytelling, but they do want you to make sure you don't forget to tell people about the problem and its solution.

This is a short clause pointing out a fairly obvious fact: you must advise the relevant parties. Whether these are parties responsible for causing a product nonconformity issue to arise, such as your suppliers, the users of your product, or the regulatory authorities overseeing it all, this is where you make sure they all know that a problem exists and that you have either solved it or are solving it.

The clause may be short and self-evident to a conscientious manufacturer, but the implementation of it may leave you breathless and exasperated already at the start line. The communication may, for example, call for user manuals to be updated (also perhaps involving a new round of usability studies), the material translated to an array of languages, and you may also need to organize user training for your customers. This all depends on the circumstances, of course, but it may be that walking the walk here is more like a marathon than just the "one more thing" Columbo – and Steve Jobs also – is famous for. Conversely, in some cases, the advice required of you may simply be placed in the release notes of your next software release for the same effect.

The relevant clauses of both standards are:

- ISO 13485 Clause 8.3 Control of nonconforming product
- ISO 13485 Clause 8.2.2 Complaint handling
- ISO 13485 Clause 8.2.3 Reporting to regulatory authorities
- IEC 62304 Clause 9.3 Advise relevant parties

In the following, we will look at the expectations of both standards, one standard at a time, and then synthesize a plan to meet the requirements of both.

9.3.1 Expectations from ISO 13485

Clause 8.3.1 expects the manufacturer's evaluation of the nonconformity to include a determination of the need for notification of any external party responsible for the nonconformity. The standard addresses nonconformities discovered before delivery and after delivery separately. No instruction on advising relevant parties is made in the case of the former, but in the case of the latter Clause 8.2.3 requires the manufacturer to have processes in place for issuing advisory notices in accordance with applicable regulations. Similarly, in the case of a complaint, Clause 8.2.2 points out that if activities outside the organization contributed to the complaint somehow relevant information will be exchanged with them. The definition of an external party ought to be understood quite broadly here.

9.3.2 Expectations from IEC 62304

Clause 9.3 succinctly points out that the manufacturer must, as appropriate, advise relevant parties of the existence of the problem. A non-binding note clarifies that the manufacturer will identify the relevant parties and that problems may be discovered before or after release, and inside or outside the manufacturer's organization.

9.3.3 Suggested Synthesis

The requirements of the two standards are complementary through and through. Table 9.4 provides a glance over the main requirements of both

Table 9.4 Synthesis of Top-Level Requirements

Combined requirements	ISO 13485	IEC 62304		
		A	B	C
Determine need for notification of any external party	✓	✓	✓	✓
Process for issuing advisory notices	✓			

standards. Note the software safety classification used by IEC 62304 and remember that some clauses of ISO 13485 may be excluded with a valid rationale (see Juuso 2022).

In practice, you will want to ensure your process instructs relevant third parties to be notified. For issuing advisory notices and informing regulatory authorities you may decide to refer to your SOP for communications (SOP-7 in Juuso 2022) instead of duplicating instructions here.

9.4 Use Change-Control Processes

Don't reinvent the wheel for implementing changes. Instead use the processes you have already defined to sort out any complaints, conformities, nonconforming product, and product changes the process for problem investigation brings up.

After you have investigated the problem arising out of a complaint, a detected nonconformity, or some discovered nonconforming product you will know what you want to fix and how. The place to do that is here and the process to follow in making the changes is something you get to pull out from under the desk as something you made earlier. All television chefs and how-to Youtubers will smile on approvingly.

The relevant clauses of both standards are:

- ISO 13485 Clause 7.3.9 Control of design and development changes
- ISO 13485 Clause 8.2.2 Complaint handling
- ISO 13485 Clause 8.3 Control of nonconforming product (incl. rework)
- ISO 13485 Clause 8.5 Improvement (incl. CAPA)
- IEC 62304 Clause 9.4 Use change control process

In the following, we will look at the expectations of both standards, one standard at a time, and then synthesize a plan to meet the requirements of both.

9.4.1 Expectations from ISO 13485

ISO 13485 provides distinct requirements for handling the sea monsters you dredged up while investigating that initial problem report. For complaints, you will have an instructed complaint-handling process (see SOP-7 in Juuso 2022). For nonconformities you will get to deploy your CAPA process (see SOP-2 in Juuso 2022), and for nonconforming products, you may have a special process that may also address the rework of already produced stock (see SOP-2 in Juuso 2022). For making changes to your software product, you will have a process stemming from product realization (see SOP-10 in Juuso 2022). You may invoke all or some of these processes based on exactly what the problem report warranted as actions.

9.4.2 Expectations from IEC 62304

Clause 9.4 requires the manufacturer to approve and implement all change requests while observing the requirements of the change-control process defined in Section 8. This somewhat awkwardly worded requirement does not, of course, mean that all change requests have to be approved, but it does mean that all approved change orders have to be implemented and not left in some limbo. It also means that problems are not fixed via some magic du jour, but via the very process you defined for changes in Section 8 – note though that any requirements for nonconformity handling and reporting to authorities from your QMS must also be observed.

9.4.3 Suggested Synthesis

The requirements of the two standards are complementary through and through. Table 9.5 provides a glance over the main requirements of both

Table 9.5 Synthesis of Top-Level Requirements

		IEC 62304		
Combined requirements	ISO 13485	A	B	C
Handle complaints, nonconformities, nonconforming product (incl. rework), and D&D changes as instructed	✓	-	-	-
Approve change orders		✓	✓	✓
Implement approved change orders		✓	✓	✓

standards. Note the software safety classification used by IEC 62304 and remember that some clauses of ISO 13485 may be excluded with a valid rationale (see Juuso 2022).

In practice, you will want to address how a complaint, a nonconformity, a nonconforming product, or a D&D change may represent a problem, and how you want the resolution of that problem to take place. The area of most interest in the case of software is most likely going to be in making D&D changes which we addressed in detail in Section 8.2.

As a final note here, don't think that bug fixes get a dedicated path flying somehow under the radar here. Klingons may have cloaking devices, Frodo may have had an invisibility cloak, and the great invisible man may be see-through under those layers of gauze, but you don't have such tricks up your sleeve here. We have often heard it argued that bugs are treated separately from everything else, that working on bugs can't possibly affect the conformity of the device negatively, and that everyone wants bugs fixed yesterday already. This may occasionally all be true, but the standards are adamant that you use your change-management process here. Bugs do not get priority access through the pre-flight security check, even if both you and your regulators are eager for you to process them quickly. Your change-management process may, of course, scale to the type of change as is appropriate based on how fundamental or superficial it is and how closely connected it is to the intended use of any medical device.

9.5 Maintain Records

Show me the money, yelled Cuba Gooding Jr. in Jerry Maguire, *and Tom Cruise did his best to repeat after him in an effort to ensure him of the promised results. In medical devices evidence always matters, and without that black-and-white it's hard to trust that something happened the way you later explain. Therefore, it's hardly surprising that you are required to generate and maintain that evidence.*

Clause 9.5 is an exercise in pointing out the obvious, but the reason it takes a Jack Nicklaus approach to driving home its message on retaining evidence is that evidence does matter and may be accidentally forgotten in the heat of the moment.

The relevant clauses of both standards are:

- ISO 13485 Clause 4.2.5 Control of records
- ISO 13485 Clause 7.3.10 D&D files
- IEC 62304 Clause 9.5 Maintain records

In the following, we will look at the expectations of both standards, one standard at a time, and then synthesize a plan to meet the requirements of both.

9.5.1 Expectations from ISO 13485

As noted above, a note that Clause 7.3.10 on D&D files instructs records on conformity to D&D requirements and D&D changes to be maintained, which is something you may be acting on here either by investigating problems or making changes. You should also make a note to use your process for control of records (see SOP-1 in Juuso 2022).

9.5.2 Expectations from IEC 62304

Clause 9.5 requires the manufacturer to maintain records of problem reports and their resolution (incl. verification). The risk management file must also be updated as appropriate.

9.5.3 Suggested Synthesis

The requirements of the two standards are complementary through and through. Table 9.6 provides a glance over the main requirements of both standards. Note the software safety classification used by IEC 62304 and

Table 9.6 Synthesis of Top-Level Requirements

Combined requirements	ISO 13485	IEC 62304		
		A	B	C
Maintain records on conformity	✓	-	-	-
Maintain records on changes	✓	✓	✓	✓
Use record control	✓	-	-	-

remember that some clauses of ISO 13485 may be excluded with a valid rationale (see Juuso 2022).

In practice, you will want to think about what records you will need from problem resolution, and ensure your SOP instructs their use in accordance with your mechanisms for control of records (see SOP-1 in Juuso 2022).

9.6 Analyze Problems for Trends

You don't have to be John Nash and see patterns in the movements of pigeons, but you must keep an eye on the encountered problems to identify points of further improvement for your product and processes.

It's not enough to fix something as you must also learn from your mistakes to hopefully prevent them from occurring again. You can employ anything from providence to artificial intelligence to analyze the data for trends, but make sure you can defend your plan to yourself, and that you have a plan.

The relevant clauses of both standards are:

- ISO 13485 Clause 8 Improvement
- IEC 62304 Clause 9.6 Analyze problems for trends

In the following, we will look at the expectations of both standards, one standard at a time, and then synthesize a plan to meet the requirements of both.

9.6.1 Expectations from ISO 13485

Clause 8.1 requires the organization to plan and implement monitoring, measurement, analysis, and improvement processes needed to a) demonstrate conformity of device, b) ensure conformity of the QMS, and c) maintain the effectiveness of the QMS. This is explained to include the use of appropriate methods such as statistical techniques.

It is technically a matter of definition whether problems are part of this monitoring and analysis, as no such direct requirement is made. In practice, claiming that problems, and the events implicated in problems, are not a matter of conformity for the product and the QMS is akin to declaring criminal insanity. Thus, the requirement in Clause 8.1 is wider than IEC 62304 problem resolution, but it should very much apply to problem resolution. Notice that the clause also requires you to strive for improvements based on the captured information and trends, not just nonconformities themselves.

The time for planning and implementing the above is before the beginning of time, and the place and time for analyzing the trends may, for example, be in the context of an internal audit or a management review. In any case, you usually do not need to analyze trends after each problem is resolved unless there's special cause to do so. For more discussion on audits and management reviews please refer to SOP-6 in our previous book (Juuso 2022).

9.6.2 Expectations from IEC 62304

Clause 9.6 requires the manufacturer to perform analysis to detect trends in problem reports. The timing of this trend analysis, or what the manufacturer should then do, is not commented on.

9.6.3 Suggested Synthesis

The requirements of the two standards are complementary through and through. Table 9.7 provides a glance over the main requirements of both standards. Note the software safety classification used by IEC 62304 and remember that some clauses of ISO 13485 may be excluded with a valid rationale (see Juuso 2022).

In practice, you should not be content with having fixed some problem presented to you. You should also look for patterns and reoccurring problems and improve your processes and operations accordingly. The intent is the same in both standards, but ISO 13485 is more vocal on the how and why as seen above. The classification and labeling of problems, the implementation of the database holding the data, and the available statistical tools as well as the long-time scales sometimes required for problems to reoccur

Table 9.7 Synthesis of Top-Level Requirements

Combined requirements	ISO 13485	IEC 62304		
		A	B	C
Monitor, measure, and analyze trends in problems	✓	✓	✓	✓
Improve based on detected trends and information	✓	-	-	-

will be factors in developing an effective process here. The choices to make are not so much governed by the standards or the regulations as by your knowledge of the subject matter, past trends, and your command of the various tools involved. There is, for example, no standard or regulation that says guest rooms must have their doors fully intact, but if you stayed as a guest of Jack Nicholson at the Overlook Hotel and started to notice doors along your corridor manifesting signs of axe blows one by one you might decide to act.

9.7 Verify Software Problem Resolution

Good planning may be half the battle, but that also means planning will only get you halfway there. Here the standards want you to ensure that whatever you saw as the remedy to the problem has been implemented, has worked, and hasn't led to new unhandled problems. Think of Bruce Willis staying back on that asteroid to ensure its detonation in "Armageddon", but don't feel verification always has to be that fatalistic.

Verification of problem resolution for software products is quite a compact clause in IEC 62304 but it has much longer, deeper roots in ISO 13485 even if the term problem is not shared across the two domains. Don't go overboard here trying to realign the whole of ISO 13485 to the single topic of problem resolution but consider how the key expectations of that standard map to what IEC 62304 wants to see. The following will take a fairly broad look at related ISO 13485 clauses, but keep your ship sighted at the expectations of IEC 62304 here.

The relevant clauses of both standards are:

- ISO 13485 Clause 4.1.3
- ISO 13485 Clause 5.5.1 Management responsibility
- ISO 13485 Clause 8.5.2e Corrective action
- ISO 13485 Clause 8.5.3d Preventive action
- IEC 62304 Clause 9.7 Verify software problem resolution

In the following, we will look at the expectations of both standards, one standard at a time, and then synthesize a plan to meet the requirements of both.

9.7.1 Expectations from ISO 13485

ISO 13485 does not talk of problems per se, but it does expect you to handle complaints (see SOP-7 in Juuso 2022), nonconformities, nonconforming product, and possible rework (see SOP-2 in Juuso 2022), as well as D&D changes (see SOP-10 in Juuso 2022) as instructed in your QMS. Invoke the correct verification activities here by referring to the instructions in the SOP in question.

ISO 13485, in general, wants to see all of your operations achieve a high level of existence where your entire quality management system runs along like clockwork. Any issues that creep up should be handled efficiently and with safety in mind, but instead of just instructing reactions to events the standard wants to see processes and resources in place in order to ensure a continuously well-oiled machine. On the top level, thus, verification of a problem-free existence for your processes and products begins with Clause 5.5.1 which instructs the top management of your organization to document and empower the personnel who verify work affecting quality. Verifying problem resolution certainly qualifies as such work. Similarly, Clause 7.5.4 discussing verification of continually met product requirements in the context of servicing activities may be seen as a related clause to ensuring the continuance of that jovial existence. Clause 4.1.3, too, may be a part of the equation by calling for monitoring, measuring, analyzing, and improving QMS processes in a way that ensures both their operation and control. Clause 8.2.6 also chimes in here by requiring monitoring and measurement of product, thereby hoping to detect any issues early on. The most reactive clauses of the set here, Clauses 8.5.2e and 8.5.3d, then call for a process to verify that the corrective and preventive actions you may have ended up implementing as a response to a detected issue don't adversely affect your conformity. All of these process links may affect your instruction for software realization here.

In case you had a problem report that then spawned nonconformities to address it will be up to you to define when that problem report should be closed: when all the child issues have been closed, or when they have been left to run their acceptable, prescribed process. The quality of your overall operations may not be best served by keeping all items open for extended spans of time, but you will want to ensure that the actions you prescribed during the investigation are carried out on the correct time scale. Even so, some actions may be affected immediately and others necessarily play out over longer timescales. The definition of done is something to consider here too.

9.7.2 Expectations from IEC 62304

Clause 9.7 requires the manufacturer to verify the resolutions made to determine whether a) the problem has been resolved (incl. the closing of the problem report), b) adverse trends have been reversed, c) change requests have been implemented in the appropriate medical device software and activities, and d) if additional problems have been introduced.

A particularly thorny requirement here is brought up by Clause 9.7b which requires adverse trends to have been reversed. Requiring this as part of the resolution of each individual problem, instead of as a part of some long-term process and product-monitoring activity is potentially a massive roadblock to a smooth problem-resolution process running predictably from start to finish. In some cases, the true assessment of the reversal may require months, or even years, worth of monitoring data to be available to capture comparable time-series data. This would lead to problem reports staying open for unjustifiably long periods of time, and each report then hogging some sliver of resources indefinitely. In some other instances, it might be possible to implement a fix, run a test sample, and see if the fix has worked or not by inspecting the devices coming off the production line. For software this is probably not going to be the case.

It is easy to see why trend analysis, or the identification of persistent problems, is an important tool. In the 1986 disaster involving the space shuttle Challenger a reoccurring fault with the O-ring seals was blamed for the loss of the spacecraft and the seven lives onboard. The issues with the seals were in fact detected long before the accident, but instead of fixing the problems it appears that ineffective, perhaps cheaper workarounds were devised (incl. limiting the weather conditions for launch, relying on the existence of a second seal). The problem was recorded in a NASA database, which in fact exasperated the issue further as it became an acknowledged, accepted point of failure that sometimes raised its head but hadn't yet stopped any flights. Had the trend been taken seriously the disaster may never have happened. Based on the story, it is easy to advocate that all problem reports should be kept open forever and repeatedly studied after the fact to detect any lingering concerns. With infinite resources, this would work, too. This might, however, cause the organization's resources to be spread thin over time, and not enough attention be given to the problems that truly were continuing concerns. Instead, it might be better to include a review of problem databases for similar issues during the initial assessment of a problem report. This way no reports are kept open unnecessarily and all of the resources can be focused on meeting the present needs as well as possible. In the Challenger example, the adverse trend could have been reversed by noticing just once that the problem had actually occurred previously instead of just trying to circumvent the problem at hand. This would not have needed an extra, after-the-fact round of analysis, just a realization of what information was available when a problem was reported.

Our suggestion here with Clause 9.7 is to assess what the reversal of trends translates to in the case of the problem and decide whether the problem report can be closed with or without some long-term follow-up monitoring. Ideally, the long-term monitoring should utilize ISO 13485 processes for monitoring and improvement, and not leave unhelpful parallel threads hanging here. Your circumstances may, however, vary and complacency can come just as easily by having a catch-all monitoring process running behind everything as with a pre-existing problem report in a database.

9.7.3 Suggested Synthesis

The requirements of the two standards are complementary through and through. Table 9.8 provides a glance over the main requirements of both standards. Note the software safety classification used by IEC 62304 and remember that some clauses of ISO 13485 may be excluded with a valid rationale (see Juuso 2022).

In practice, all of the requirements made by IEC 62304 are in line with the expectations of ISO 13485, but the same sentiments are addressed under

Table 9.8 Synthesis of Top-Level Requirements

Combined requirements	ISO 13485	IEC 62304		
		A	B	C
Observe your processes for complaints, nonconformities, nonconforming product (incl. rework), and D&D changes	✓	-	-	-
Verify that the problem has been resolved and the problem report closed		✓	✓	✓
Verify that adverse trends have been reversed		✓	✓	✓
Verify that change requests have been implemented in the appropriate medical-device software and activities		✓	✓	✓
Verify that additional problems have not been introduced (or are handled appropriately)		✓	✓	✓

the discussion of, for example, complaints and nonconformities, instead of an entity called a problem. You will want to ensure that the above IEC 62304 requirements are met by your processes when they concern a problem with your medical-device software product or its manufacturing. This also includes programs, procedures, and associated documentation and data, as pointed out by IEC 62304.

9.8 Test Documentation Contents

When the stuff has somehow hit the fan, and you are eager to get the fix out, it may be tempting to forget about testing the fix in a documented manner. Yet, the standards don't want to see you forget about retaining the necessary information.

Testing as part of the software-development processes was covered in Clause 5 (see Section 5.7, for example). Here the focus is on testing as it relates to detected problems, their verified resolutions, and the documentation to retain. By now you will have noticed that IEC 62304 is quite big on testing and test records, which incidentally form the core of the documentation needed also for legacy software. It is therefore not surprising that test documentation gets its own short section under problem resolution.

The relevant clauses of both standards are:

- ISO 13485 Clause 4.2.4 Control of documents
- ISO 13485 Clause 7.1 Planning of product realization
- ISO 13485 Clause 7.5.8 Identification
- ISO 13485 Clause 8.2.6 Monitoring and measuring of product
- IEC 62304 Clause 9.8 Test documentation contents

In the following, we will look at the expectations of both standards, one standard at a time, and then synthesize a plan to meet the requirements of both.

9.8.1 Expectations from ISO 13485

Clause 7.1 requires you to plan testing in planning product-realization. Here, in this section, we are talking about reacting to detected problems and testing after a fix has been applied instead of planning a new product-realization process, but nonetheless your SOPs should instruct problem resolution, and, for example, your software development plan or maintenance plan may then expand on the arrangements and resources involved. Thus, Clause 7.1 may indirectly apply here.

The impact of Clause 7.5.8 is more direct as it requires the identification of product status to be maintained throughout product realization to ensure that only product having passed the required inspections and tests, or released via an authorized concession, is put out. Clause 8.2.6 points out that if test equipment is involved in the monitoring and measurement of products the records must identify the equipment as appropriate, and, for implantable devices, also the testing personnel must be identified (Clause 8.2.6).

Finally, note that Clause 4.2.4 requires documents to which medical devices have been tested to be available for at least the lifetime of the device, but not less than the retention period for any resulting record, or as dictated by applicable regulatory requirements. You will have defined the retention period and device lifetime in your QMS (see Juuso 2022), but here make sure that the documentation you produce meets this requirement. The documentation you produce here should be classified as records, while the instructions for creating those records will be in your documents (e.g., quality manual and SOPs).

9.8.2 Expectations from IEC 62304

The final clause of the standard, Clause 9.8, addresses what your test documentation for changes must include. The testing itself is explained to take the form of testing, retesting, or regression of software items and systems following a change. The instructed set of information here is as follows: a) test results, b) anomalies found, c) version information of the software tested, d) hardware and software configurations, e) test tools, f) testing date, and g) identity of the tester. Items d) and e) are qualified with the call for relevant information to be entered, and thus the answers to these fields may be brief or verbose depending on your circumstances.

9.8.3 Suggested Synthesis

The requirements of the two standards are complementary through and through. Table 9.9 provides a glance over the main requirements of both standards. Note the software safety classification used by IEC 62304 and

Table 9.9 Synthesis of Top-Level Requirements

Combined requirements	ISO 13485	IEC 62304		
		A	B	C
Ensure appropriate control of documents and records	✓	-	-	-
Record test results	✓	✓	✓	✓
Record anomalies found	-	✓	✓	✓
Record version information of the software tested	-	✓	✓	✓
Record relevant hardware and software configurations	-	✓	✓	✓
Record relevant test tools	✓	✓	✓	✓
Record testing date	-	✓	✓	✓
Record identity of the tester	✓	✓	✓	✓

remember that some clauses of ISO 13485 may be excluded with a valid rationale (see Juuso 2022).

IEC 62304 is more specific in its requirements here but in practice, all of the expectations are the same between the standards. ISO 13485 expects evidence to be on record for proving any activity and it places great importance on the immutability and auditability of that information. Thus dates, identities of testers, and details of the testing environment become very important for repeatability and traceability reasons. In practice, IEC 62304 only adds a few software-specific requirements.

9.9 The Parts Left Out by IEC 62304

In general, problem resolution for software products is a topic for IEC 62304 and is not covered as such in ISO 13485. However, considering the many incarnations a problem can have in the QMS, the product-realization processes defined within it, and the product itself, problem resolution does have much longer, deeper roots in ISO 13485. A problem detected in the IEC 62304 sense may have implications that, for example, take the form of a complaint, nonconforming products, or a nonconformity in the processes. All of these must be addressed in the appropriate ways meeting the expectations of both standards.

Don't go overboard here trying to realign the whole of ISO 13485 to the single topic of software problem resolution but consider how the key expectations of that standard map to what IEC 62304 wants to see. In the above, we took a fairly broad look at the related ISO 13485 clauses, but here at the end of Clause 9 we thankfully find that all the key issues and requirements from both standards have now been addressed.

Chapter 10

INTEGRATION WITH YOUR QMS

Phew! That was a massive amount of detail on your software product-realization activities dished out from two perpendicular points of view. Each standard makes perfect sense by itself, but to marry the two together and end up with something that makes sense in a lean, practical way takes real effort. This is where we will start to put those notes together and shape that symphony you need.

The master plan set out in this book is to negotiate together the requirements of the two foundational standards, ISO 13485 on quality management and IEC 62304 on software development, in a way that has you meeting all of the requirements but not adding unnecessary hoops to jump through. ISO 14971 risk management, too, is important, but in our task here it can be treated as an opinionated aunt popping by every once in a while. In practice, risk management cuts through everything, and offers an alternate, perpendicular view to what we are constructing here. This means that we can, and should, start with the two other standards here to build our network of processes. As the annexes to IEC 62304 are quick to point out, you may already have a QMS in place prior to adopting the software life-cycle standard. In this case, you will naturally perform some sort of a gap analysis (e.g., internal assessment, internal audit) to assess your starting point first, and then act as appropriate to align your processes with the discussion in this book.

The key underlying notion here is that the foundations of your operations are described by ISO 13485 quality management. The QMS you have built according to this standard will give you the basic toolbox to run your organization and ensure that the products you place on the market meet their requirements. This includes, among other activities, making sure that the review, verification, and validation activities you run meet the expectations set by the standard. In terms of the reviews and verifications you conduct during software development, the meat will be in IEC 62304, but the packaging required by ISO 13485 will help ensure that important points of view are observed. The same is true for planning and requirements analysis also, and especially your validation activities which are not covered by IEC 62304 at all.

Thus, forging a cohesive software-development pipeline with minimal repetition, back-and-forth, and unnecessary overhead comes down to negotiating the following pieces of the puzzle.

- **Standard operating procedures (SOP)**
  The standard operating procedures of your QMS that describe your product realization process and the other supporting processes.
- **Planning documentation**
  The planning documentation for your software-development project, incl. the software development plan (SDP).
- **Execution of the project**
  The execution of your development project according to your documented arrangements (incl. the creation of the set deliverables).
- **Execution of the post-release activities**
  The execution of your post-release activities.
- **Continuous improvement of the QMS**
  The continued improvement of your QMS.

If you succeed in addressing all of the above in your software-development project, you will meet the requirements of both standards. Each of the above pieces of the solution is discussed in a section of its own in the following.

Note that Annex D in IEC 62304 attempts to give a rough checklist for the adoption of the standard by small manufacturers who do not yet have a certified QMS in place. In our view, this checklist is not all that helpful as it boils down to a table of contents for the two main standards, but it is worth knowing about its existence. In any case, software development according to IEC 62304 can not survive for long, if at all, without ISO 13485 quality management and ISO 14971 risk management. Annex D does not change this basic assumption of the standards.

10.1 Write Your Product Realization Processes in the QMS

This book has walked you to the edge of the reservation and pointed out the important markers you should observe on your trek forward. However, there is no magical shortcut to now putting one foot in front of the other, or pen to paper, and plotting your course. Let this book be your Yoda on the road ahead, but don't expect the finished QMS to appear as a vision by itself.

You must now make sure your processes match the expectations of both IEC 62304 and ISO 13485 – and any regulations applicable to you. The discussion in this book, and that given in our previous book focusing on ISO 13485 QMS by itself (Juuso 2022), will help you in writing the necessary processes one by one. Take the instructions given on the QMS as your foundation here. Then add all the detail expected by IEC 62304 under the specific software development (see Section 5), maintenance (see Section 6), configuration management (see Section 8), and problem-resolution processes (see Section 9) under product realization (see SOP-10 in Juuso 2022). Don't forget about risk management either (see Section 7).

You should also appreciate the influence of the customs and habits your organization may have adopted over past software-development work, and the properties of your chosen software-development model. These will weigh in here as you bend the requirements to your will – or, perhaps, the other way around.

In writing the SOPs also consider the documentation you will want to create as part of performing work according to those processes. This will include many of the various plans, reports, reviews, and other types of records your SOPs want to conjure into being. You will want to create templates for some of these records, and thereby perhaps reduce the amount of prose needed in the SOPs to list all of the relevant properties and fields of content that are required for each. As for documents, we recommend that you stick to defining documents as your top-level QMS documents: your quality manual, your SOPs, and perhaps your working instructions. See SOP-1 in our previous book (Juuso 2022) for more discussion on documents and records.

Perhaps the greatest trick to designing a modern software-development process while meeting the requirements of the standards is how you marry your agile development cycle with the waterfall-like expectations of the standards – in other words, how you turn that classic waterfall into a modern washing machine with the spin cycle your organization expects to have at its disposal. Table 10.1 below sketches out one possible solution. Here the overall structure consisting of four phases matches with the waterfall model of the standards. The typical development tasks of planning, requirements elicitation, design, implementation, testing, release, and maintenance

Table 10.1 Basic Software Development Life Cycle

#	Phase		Key Deliverables
1	Product-Realization Planning		Software Development Plan (SDP) System Requirements Risk Management Plan
DIR	**Design Input Review**		**Review Minutes**
2	**Sprint Planning**		Sprint Management Records
	Repeated incrementally	Requirements analysis	Software Requirements Clinical Evaluation Plan (CEP) Risk Management File (RMF)
		Design	Software Architecture Design [B][C] Software Detailed Design [B][C] Test Cases (incl. plan and procedures)
		Implementation	Software code
		Testing	Software Verification Report Software Validation Report Software System Test Report
	Sprint Review		Sprint management records
DOR	**Design Output Review**		**Review Minutes**
3	Release		Software Maintenance Plan Instructions for Use (IFU) and Label Documentation of Residual Anomalies [B][C] Risk Management File (RMF) Clinical Evaluation Report (CER) Post-Market Surveillance Plan Draft Declaration of Conformity
DTR	**Design Transfer Review**		**Review Minutes** Final Declaration of Conformity
4	Maintenance		Post-Market Surveillance Report or Periodic Safety Update Report (PSUR) Post-Market Clinical Follow-up Report (PMCF report)

are punctuated by the three major reviews discussed earlier in this book: DIR, DOR, and DTR. Given a set of requirements at the top of the waterfall you could let gravity take over and end up with a finished piece of software downstream. The magic of turning that age-old into something more agile takes place in the second phase of development as depicted in Table 10.1.

Table 10.1 sketches out a four-phase approach to meeting both industry-standard expectations for agile software development and the requirements of the medical-device standards and regulations at the same time. The motivation for the model here has been to allow for the typical sprint-based approach to development while ensuring that the expected analysis and reviews take place at the appropriate steps. The simple idea behind the model is that the key documentation listed in the right-side column is available for each subsequent review, and once all of these can be approved you may move to the next phase. Note that in phase 2, the four types of activities – requirements analysis, design, implementation, and testing – are performed iteratively. You may thus first run entire sprints to flesh out the software requirements and design, and only after successive sprints finally have an implementation to test. It is thus not required to perform activities of all four types in every sprint and it is not required to finish all activities of one type to proceed onto another type of activity. In practice, you may choose a set of requirements for each sprint, use that sprint to work on their realization, and then tie the full set of requirements (and their realization) together in the design output review (DOR). Developers are used to tickets and working on a few items of the whole at a time, but the standards want to see you have a firm grip also on the whole that those items make, as discussed in Sections 5 through 9. Designing this model here is where you get to overlay the expectation of the regulated medical-device environment over the freedom software developers crave, and hopefully do so in a way that doesn't have the developers or the auditors fuming out of their ears.

Following the model introduced here you won't be attempting to do everything at the same time, you won't be jumping too far ahead of yourself, or waiting unnecessarily on some documentation that is actually needed much later in the process. This is, of course, a rough overview of the overall development process and you may choose to structure your development life cycle slightly differently in your circumstances, but overall we feel this generalization makes much sense.

In reality you will be updating a number of documents repeatedly along the way, but exactly how often you do so will depend on how good your initial documents are, and how you identify needs to update them.

10.2 Develop Your Planning Documentation According to Your QMS

After you have instructed your processes in agreement with the standards, ensure that they are also in agreement with your everyday practice. In other words, it is not enough to write a beautiful, leather-bound work of fiction for your QMS if you don't then live by its tenets. Your SOPs should be clear enough and comprehensive enough so that you – and everyone else at your organization – can now follow them, perform their work as instructed, and create the records requested.

Sections 5 through 9 of this book will introduce you to the required records, and what is expected of each. The first record to consider here is the software development plan (or system development plan) discussed in Section 5.1. Also remember that your overall QMS will instruct other records to consider here as well, including records on software tool validation, supplier evaluation, and perhaps personnel qualification. You may not need all such records at the start of preparing your planning documentation, but these may soon become relevant as you execute that plan. Dodoism, or the act of burying your head in the sand and hoping for the best, has no place in any of these activities.

10.3 Execute on Your Plans as Expected by Your Arrangements

The only thing left is to do the thing itself? Not quite, as in the world of medical devices you are always required to also document the thing you did. Nonetheless here is where you take off those training wheels – or the unnecessary bandage around your injured arm if you are Chuck Norris – and jump into action. Section 11 next should give you a good overview of what records are expected of that action.

At the end of this phase, you will have developed your piece of software, scrutinized it adequately, and even released the final product into the world beyond your doors. You will have the finished product, all of its accompanying documentation and supporting services, and you will have documentation covering its whole birth available at will. This documentation may be requested by your notified body, regulatory authorities, or even your B2B customer.

10.4 Execute on Your Post Release Activities

This is where you are Janine Melnitz hooking up the telephone line at the brand-new Ghostbusters headquarters to open up for business, and hope that no ghouls come over the telephone wires. In the past, it might have been enough to just wait and listen, but today you are also required to actively check up on how that software you created is fairing in the real world. Your QMS will have defined a whole post-market surveillance process for this (see Juuso 2022), and here we have added detailed processes for addressing maintenance, changes, and any encountered problems. Once again what remains is to follow through on those processes.

10.5 Keep Your QMS in Good Shape

Your company is up and running, and your product is changing the world for the better. You might be thinking: why rock the boat? Maybe you shouldn't, but you have to be on the lookout for ways to improve your QMS.

ISO 13485 expects to see you embrace continuous improvement for all your operations. Change for the sake of change is not the aim here, but instead charting your position and considering ways to improve the status quo always is. This topic is addressed in our previous book (Juuso 2022), and it is not specific to software but rather quality management in the context of all medical devices. Here the strive for continuous improvement is brought up purely as a kind of heads up to not sit on your laurels after software release. In practice, a lot of the improvement activities around your QMS will be led by functions other than software development at your organization, but some of the best ideas for improvement may originate from within the very function that first gave birth to the software. The notion of the factory floor, or going right to the source, is an essential concept in quality management – even if the father of modern quality management, W. E. Deming, famously attributed 95% of all quality issues to management. If you are in management take that to heart, but if you are not, forget we said it.

Chapter 11

TECHNICAL DOCUMENTATION

Did you skip ahead to here? We wouldn't blame you if you did, but after taking a dip here remember to go back and read all the detail in the previous sections too. Developing technical documentation in a lean and practical way that matches seamlessly with your chosen life cycle model is a persistent engineering topic in the development of medical devices today. It is perhaps not as flaming-hot as leveraging new emerging technologies such as artificial intelligence, but it may have some equally profound implications for the safety and efficiency of your devices. The Holy Grail is a documentation model that adds no overhead and meets with no friction while inching its cuddly tentacles everywhere and at all times. That may not be achievable, but you will get close enough if you fashion a development model that neglects neither code nor documentation.

A stereotypical way to approach the medical-software business is to first think about the code you want to write, the documentation you have to create for it, the quality management system you have to have in place, and the cool green cash you hope to make from it all. On this path, you may think you have coding covered and the rest is just Styrofoam in the box. Not so.

As remarked on in Section 1.13, documentation is intended to be an accurate reflection of the whole that is your medical-grade software. The documentation must show what requirements the software is based on, what design choices these led to, and how those were then implemented, verified, validated, approved, and released. The risk management file must then reflect this thread from beginning to end. The goal is that instead of trying or retesting your software over an extensive period of time, the relevant stakeholder, such as your auditor, can look into the wheelhouse by referring to your documentation where they will see how it is all going. The basic tenet of IEC 62304 is that mere testing is not enough to ensure safe medical software. Instead, you must impose control over the entire development and life cycle of that software. The technical documentation should thus shadow your software as closely as possible, and if it falls behind or out of touch with the code itself somehow, at any point, you may be in trouble. If, however, both the code and the documentation match nicely, and meet the applicable requirements, your house should be in good order.

The discussion in the earlier sections of this book has identified the key types of documentation you will want to have whether or not you call the documents and records by the exact same names or not. This section will go over the expectations from both standards that are specific to your body of technical documentation, and then slot in the various pieces of documentation identified earlier to build a comprehensive and practical set of key documentation. This set will then help you answer the needs your stakeholders have both now and over time as your software evolves. Sections 5 through 9 of this book have already introduced the requirements for the individual documents and records in detail, so here we will concentrate on the overall architecture of the documentation.

11.1 What Is Technical Documentation Anyway?

Let's first start by briefly considering the terminology involved in the definition of technical documentation. The terminology is simple, but some confusion appears to exist among manufacturers, and partly this is caused by a subtle difference in EU and US terminology. In general, the terms used are as follows.

- **Medical Device File (DMF)**

    This is the documentation set expected by ISO 13485 Clause 4.2.3 (see Juuso 2022). This set of information is kept for each medical device or medical-device family and is expected to contain all necessary information for demonstrating conformity with the standard and any applicable regulations. In practice, this includes the intended use/purpose, general description, labeling (incl. IFU), and specifications for the product and its realization.

- **Design and Development Files**

    This is the documentation set expected by ISO 13485 Clause 7.3.10 (see Juuso 2022). This set of information is kept for each medical device or medical-device family and is expected to contain records demonstrating conformity to D&D requirements, as well as any D&D changes made. The implication here is that you need to maintain documentation demonstrating conformity at the start, but also to ensure any changes are reflected regardless of whether these take place a little later or a lot later.

- **Device Master Record (DMR)**

    This is a US term roughly corresponding with DMF in ISO 13485. This refers to the set of documentation kept for a specific type of device. The Code of Federal Regulations (820.181) sets requirements for the contents of the documentation which are very much in line with the expectations of ISO 13485.

- **Device History Record (DHR)**

    This is a US term for documentation kept during the manufacturing of a device on the level of batches, lots, or units. The Code of Federal Regulations (820.184) sets requirements for the contents of the documentation along the lines of manufacturing dates and quantities, released quantities, acceptance records (in relation to the DMR), and traceability.

- **Quality System Record**

    This is a US term for documentation expected by the QMS but not necessarily specific to any medical device. The Code of Federal Regulations (820.186) sets requirements for the contents of the documentation. Note that complaint files are specifically called out by US CFR 820.198.

- **Technical Documentation (TD)**

    This is the general term used in the EU for technical documentation demonstrating the conformity of a device (or device family) to the medical-device regulation. EU 2017/745 Annex II provides a detailed list of the type of documentation expected as well as the overall hierarchy it is expected to be made available under.

In practice, the above requirements can all be answered in the case of medical software via documentation corresponding to the expectations of the IEC 62304 and ISO 13485 standards. ISO 14971 risk-management documentation will have a part in this as well. In practice, you are expected to maintain relevant product and production information in a manner facilitating good traceability between D&D inputs and outputs, and over the generations, batches, lots, and units of devices you manufacture.

The information expected for your technical documentation will depend on the type of medical device you have, what regional regulations are applicable to your device, and which standards beyond ISO 13485 are applicable to your activities. The documentation will likely also include subsets such as the risk management file, the usability engineering file, and clinical evidence for your device. It is up to you as the manufacturer to marry all of these requirements together into a cohesive, navigable body of documentation meeting all of the various stakeholder needs.

Figure 11.1 shows an example of a high-level structure for technical documents. The exact nature of documentation in your set will depend on your type of device (e.g., hardware, software, or both), but the concept of providing a starting index is universal. This index will then quote and reference

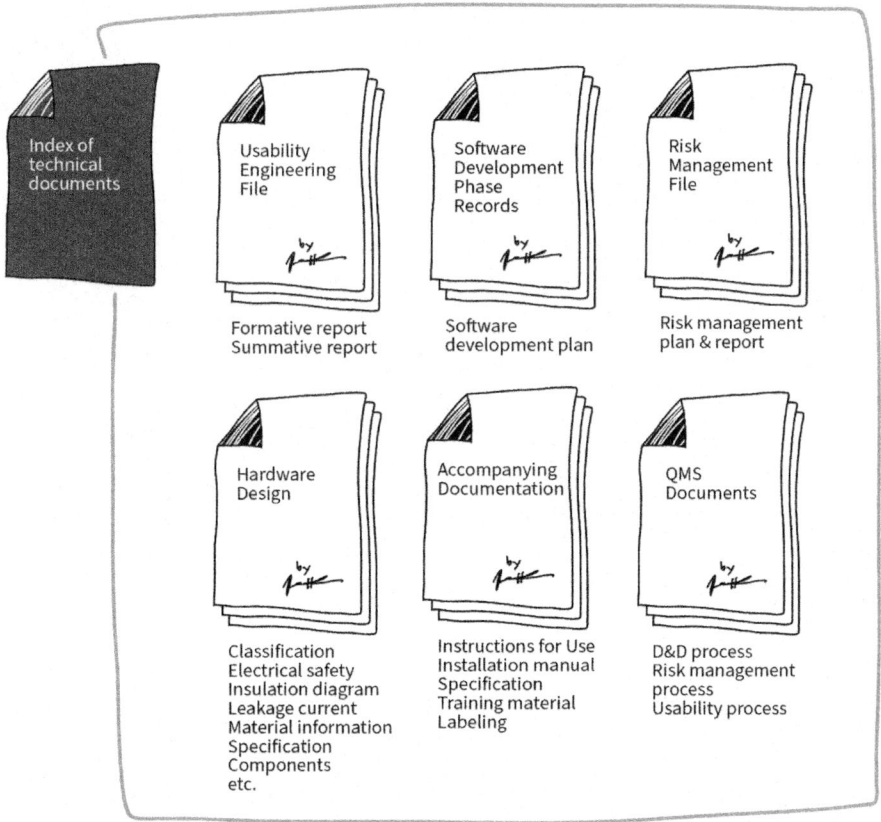

Figure 11.1 **Example structure for technical documentation.**

other documents in your set as is needed to answer questions regarding your medical device or medical-device family.

11.2 Process Documents

Much of the information listed above refers to what ISO 13485 understands as records, i.e., records retained of work having taken place, but occasionally you may also need to produce the documents instructing that work. These are the documents, as ISO 13485 speaks of them, which set up your processes through your quality manual, standard operating procedures, and perhaps your working instructions. These documents may be thought of as a part of your technical documentation, or as something existing on a layer above it, but nonetheless, you are required to retain such documents for at least the life of your records and produce them for reference when your records are being assessed.

11.3 Development Records

Given the existence of separate process instructions, the objective of planning documentation is to add any necessary further detail on the execution of those processes. After planning, all that remains, in theory, is to execute those plans, processes, and other documented arrangements to obtain the deliverables you expect. From a business point of view, here you are running your processes and doing the things you expected to do during software development. From a regulatory point of view, you are performing the planned activities, developing the deliverables, and collecting the required evidence on the quality of this all.

In Section 10 we discussed a four-phase life cycle model for software development. Table 10.1 (given in Section 10) sketched out the different phases of development work, married the work with an agile approach, listed typical deliverables, and presented the three principal reviews (DIR, DOR, DTR) used to ensure the outcomes meet expectations. The two figures below, Figure 11.2 and Figure 11.3 provide a simplified view of the technical documentation of most concern in each of the three reviews. The set of documentation sketched out by the two figures is what the standards expect from you in a more-or-less chronological order, and it is also the set

Technical Documentation ■ 257

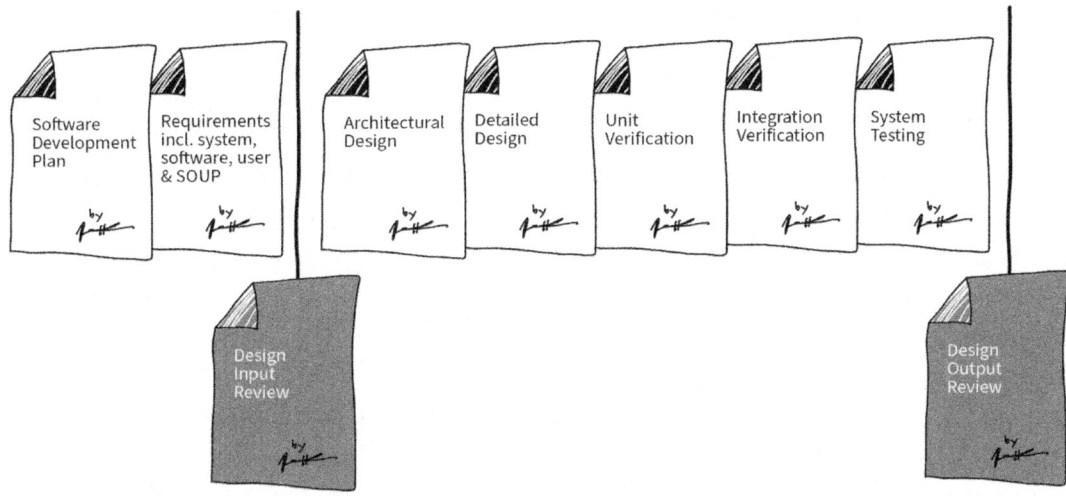

Figure 11.2 Technical documentation in DIR and DOR.

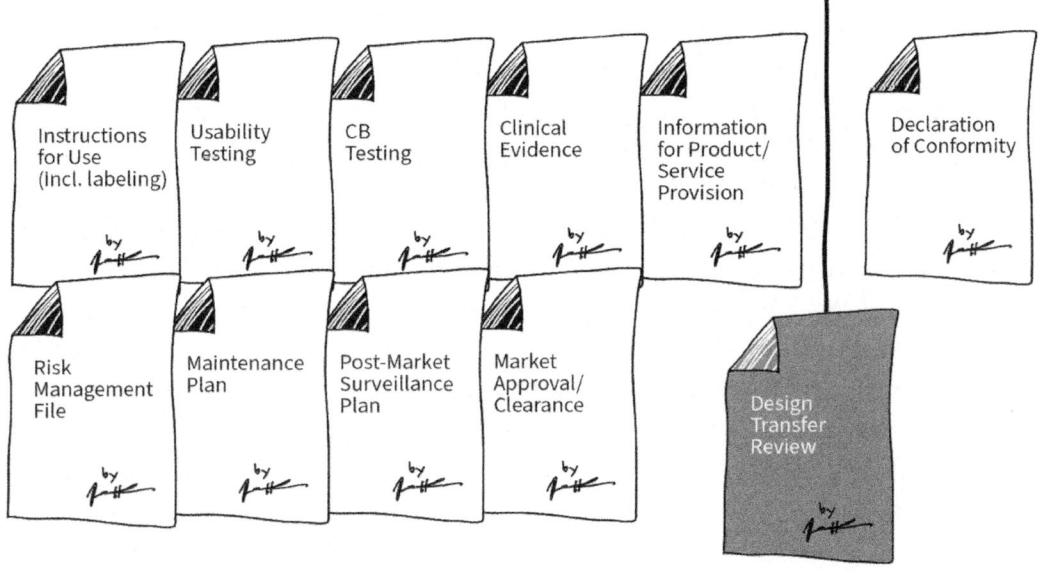

Figure 11.3 Technical documentation in DTR.

that anyone inspecting your operations from the outside in will expect to find.

Figure 11.2 illustrates what records are typically viewed during a design input review (DIR) and a design output review (DOR). In practice, you may find that seldom are all of the requirements known at the start of a development project, and these thus need to be updated perhaps throughout the project. Similarly, risk management will evolve as the project matures. Nonetheless, the figure here provides an overview of the ideal run of things.

The design documentation (incl. architectural design, and detailed design) may be developed in sequence using a waterfall-based approach or something more akin to agile methodology. The documentation must, though, make sense throughout, on both the level of the system as a whole and the individual software items that it comprises. Similarly, the body of verification activities may be built up of smaller, incremental activities that address both the components and their sum adequately. System testing and validation activities must ensure that the sum of it all meets with what both you and your customer expect to get at the end of it all.

Note that in addition to verification, a validation plan and a report are also expected sooner or later. Requirements for both are given by the IEC 82304-1 standard on health software (see Section 12.2). You may decide to have the plan available for DOR, but what is actually required by the standards is that the plan is established before validation readiness can be achieved (IEC 82304-1 Clause 6.2a), and thus before validation is carried out.

Figure 11.3 picks up after DOR and takes us to the finish line where a design transfer review (DTR) takes us to the market.

The most important of the three reviews is the design transfer review (DTR). Here you must be satisfied with launching the thing you created into the real world. You must make sure you are happy with all of your principal records, have kept them up to date, and have moved ahead in a rational, defensible manner. If there's any work to redo this is the place to notice it and act accordingly. That said, remember that the standards take a sacred view of defining what is acceptable before you actually see what you got, mostly so that you avoid the temptation of saying what you got was actually what you wanted all along. No clairvoyance is called for, but you must know what is good enough before getting there, and the place for the final check on whether what you got is good enough is here before moving to production.

After you know that you are good to go, and have received any necessary approvals or clearances from your notified body and regulatory authorities, you may finally sign the declaration of conformity (DoC) and step into the market. You will, of course, then be running your maintenance, change management, and post-market activities as promised – and churning out the records you want to obtain from that work.

11.4 The Audit Package

Given the above broad definitions of what exactly is technical documentation, it may seem that your auditors will want to see it all. They may, in fact, want to see it all, but you will probably not photocopy it all and messenger it over. Instead, you are typically asked to provide a top-level index of your technical documentation and some specific documents answering specific topics.

The EU 2017/745 Annex II referenced above gives one possible way to structure the index file for your technical documentation. Given this model, the information discussed above might, for example, take the form of the following documentation list:

- Device description and specification (incl. intended use/purpose)
- Information to be supplied by the manufacturer (incl. IFU, labeling)
- Design and manufacturing information (incl. IEC 62304 development documentation, and specifications/procedures for product realization)
- General safety and performance requirements (GSPR), as applicable
- Risk-management documentation (incl. benefit–risk analysis, significant residual risks, and anomalies)
- Clinical evidence

Much of the information listed above refers to what ISO 13485 understands as records, i.e., the outcomes of work, but occasionally you may also need to produce the documents instructing that work. Your index might therefore reference both the instructions you give for work and the individual main records kept of that work. All of your records should be under your record control as instructed by your QMS (see Juuso 2022), but you do not have to list all of your records in your index, and you do not have to ship copies of the entire body of records or provide your auditors with remote access to all of your records. You will need to produce any documents and records requested by your auditors, though. You should also expect good traceability, and the ability to move between individual records and documents in any fashion facilitating an investigation of your instructions and actions, to be required.

Chapter 12

SEEING INTO THE FUTURE

Being able to predict the plot points in films is occasionally fun, but usually just a drag chipping away at the magic of cinema. Real life is seldom about the suspension of disbelief or going along for the ride, so having some inkling of what is to come ahead of time just makes sense and lowers your blood pressure. This section will discuss the future challenges and opportunities for developing medical-grade software. Some of these topics may be immediately relevant to your type of software, while others may shape the standardization and regulatory environment around them.

Ever since the microprocessor first revolutionized what medical devices could do, we have been on a path of constantly improving upon what is possible with software. The arrival of network connections between devices decades ago has now given way to new levels of connectivity with cloud-based services and hub-centric thinking first connecting hospital wards to a central control room and then entire hospitals of different sizes together. Most recently, the application of artificial technology has heightened the need for safe solutions to understanding and controlling advanced technologies that are starting to take over tasks previously only possible for humans to perform. Humans are in no danger of being replaced as the experts in the loop – or indeed the patients themselves – but we are now getting to a point where a small part of the doctor could be thought of as embedded into the medical devices, e.g., in the form of AI-based analytics.

All this activity means that the standards and regulations, too, are under a constant need to stay current with the times. The present version of IEC 62304 is outdated, but even a new version published today would not automatically stay current for any extended period of time. For this reason, all standards have a routine maintenance cycle: all the key medical-device standards discussed in this book are revised every five years by the standardization committees responsible for them. The committees don't just meet every five years; instead many of them run constant monitoring activities and forward-looking studies into what challenges might be ahead, and how to best respond to such challenges.

This section looks at a few of the most current topics relating to the development and maintenance of medical software we see being discussed across the world. The topics discussed here are perhaps only implicitly addressed by the present generation of key medical-software standards, but they may have a profound impact on the software products of tomorrow. A few of the topics, usability engineering in particular, are ones that are intuitively easy to approach but notoriously hard to master. Other topics, such as edge intelligence, are novel new ideas that may grease the wheels in the right places to unlock stalemates or catalyze new solutions to long-standing problems such as data transfer, cybersecurity, the protection of personal information, and also intellectual property.

12.1 Agile Software Development

> *In the film* Armageddon *a group of Texan oil drillers turn everything they have learned upside down to go blow up a gigantic asteroid heading for Earth. No one would call this bunch nimble, but as an exercise in agility, the move from an offshore oil rig to a space station in a matter of weeks is worthy of entry into Cirque du Soleil. The same sort of dexterity, nimbleness, and speedy project execution is what we are talking about when considering the adoption of agile methods in medical-software development.*

This entire book has been written in a way that is as compatible with agile software development as possible, yet we felt that a separate discussion of the methodology itself in the context of medical devices is called for. That is what this section aims to achieve.

We have been following the standardization of agile software development for regulated industries over the past several years, including as members of select international standardization committees and mirror groups. Much work has been done on the standardization of agile, scrum, regops, and many other related domains, but as of yet no single standout standard for an agile approach to medical-device development has emerged. To add fuel to the fire, Dr. Shuren, the director of FDA's CDRH even remarked recently at an industry forum that agile regulations might also be a thing. This may or may not have been a slip of the tongue, but we do not yet live in a world where agile nimbleness has outsmarted the age-old waterfall.

It's a cliché to say that developers want to run free and regulators want them to sit quietly in the corner churning out perfect code to everyone's benefit. The agile world of the incremental, occasionally reactionary development work revolving around tickets and sprints clashes with the perfectly predictable, tediously planned document-centricity of the regulatory context. The answer is neither anarchy nor paralysis, but something in between. The Holy Grail here is a frictionless QMS and a maglev-levitated development pipeline that accelerates any idea to a piece of safe software in less time than it takes for the subsequent money orders to go through in the opposite direction. Negotiating the everyday requirements of software development in a business setting with the sometimes lofty expectations of the standards and regulations is, however, a task to reckon with. In a world without mistakes, misinterpretations, imperfection, and friction this might be an easy task, but in this one, and in a domain as critical as medical devices, a good solution here takes real effort. The good news is that this goal is routinely attempted, and passable or even good solutions are achieved. Our hope in this book has been to point out the important design considerations involved and even provide some of the smart solutions you might employ. Nonetheless, agility and the safe mastery of agile methods in the medical setting is a mission greater in scope than any book or standard written on the topic. It is a lifelong undertaking, but one hopefully revolving around the sort of considerations developed in this book.

12.1.1 What Is Agile Anyway?

Wizardry in a vacuum, the lone code guru as a prophet of their profession, is not such a hot idea when the objective is to consistently create code that fulfills some real need the users have. Error-free code doesn't just appear by itself, and a black-box approach to writing software even when you use intelligence of the natural kind – and don't just rely on AI or ChatGPT to write your code – is a risky proposition. If you like hanging around in your underwear all day and constantly sliding from one task to the next like Tom Cruise in that famous scene from his non-ISO-14971 business film, you may decide that lifestyle is for you. For the rest of us, there are standards and software-development life cycles to choose from.

The first development model that early mankind mastered after learning how to spark code may have been the waterfall model. Even then man was thinking about the kinetics of the maglev development pipeline we just brought up: if you push a boat down a series of waterfalls and it makes it through all the trials you may have something worth putting out. If that waterfall is Niagara Falls, and the thing you built travels better than a wooden barrel you may even have ended up saving some daredevil lives a hundred years ago. In any case, gravity worked just like magnetic forces to remove any friction from the development path.

Today the waterfall model is recognized as a simple concept to grasp, even a practical framework to use when describing requirements for development work in standards. The model provides handles on the wizardry and controls over its outcome, which is why it has been successfully used at least since the 1950s. The model has its origins in the manufacturing and construction industries and is generally well-suited for the project management of undertakings where each subsequent step relies on the deliverables of the previous steps. Think of *Tetris* with a greatly expanded field of vision. If, however, the falling blocks start to get flakier and their specifications may evolve in some incremental manner as they fall down your screen you might want to reach for something a little faster-acting than the waterfall model. Think of Schrodinger's cat and how much more fun *Tetris* would be if the blocks took after his offspring.

The waterfall model was good for setting checkpoints on the work performed and setting criteria for moving ahead to the next phase. This, no doubt, opened enough eyes and provided enough checks so that errors were greatly reduced from any past state of the art. Viewed from a modern lens, though, the waterfall model has started to seem too unresponsive and also out of touch with the everyday of software-development work taking place. A model that is viewed as a drain on resources and one which no longer provides the required timely insight into the work and control over it might need refreshing.

To put the granny on wheels, a new agile software-development life cycle was developed. Not having to plan each possible path forward in the *Marvel* multiverse of infinite possibilities, and at the same time retaining a better understanding of the path that we actually were on at the moment, probably led to the coining of the principal agile models we have today. Part of the solution here was to narrow down the intervals between checkpoints so that we might check up on the progress more frequently and take smaller steps toward the right end goal. In other words, by using an agile approach we would be building up the capabilities of the software constantly and with more certainty of heading in the right direction. Now we wouldn't just have to wait for the development to reach its next stage, but we could better remain in touch with the work that was going on all the time.

The advantage of agile software-development models (e.g., XP, scrum, kanban) is the increased velocity of implementing new features and functionality for software. This, in theory, decreases the total expenses involved and improves the transparency of the throughput. In a regulated industry such as medical devices, however, velocity alone does not rule the day. Here, the availability of a new feature takes a second seat to the reliability of that feature as the stakes may be life and death. It thus becomes extremely important to observe the standards and applicable regulations, which in turn call for truly understanding the users, the risks involved, and the benefits sought. Incorporating these aspects of medical device development appropriately into the agile model may feel like it sucks out the benefits of the model.

The inclusion of the various planning, review, verification, validation, and risk-management activities expected by regulations may feel like extra work during development. These are a natural part of the waterfall model, but as we start sprinting in cycles it may start to feel repetitive and unnecessary to see to these activities each time, all the time. Indeed, performing and reperforming all of this for every sprint will be unnecessary, but forgetting to perform them at all or in the right place in time will be unforgivable. Often times the trick here will be to allow for the incremental, piece-wise working model of agile, but also remembering to adequately address the whole that is built from those incremental steps. A case in point is given by the requirements definition: here you may be tempted to hone in on the final set of requirements over time, but don't forget that you must also ensure the set of requirements makes sense as a whole, covers all of the bases it is required to, and none of the requirements contradict one another. This sounds simple enough, but as you are adding to and removing from your set of requirements over sprints this will start to feel less simple. The more such loose ends you find yourself leaving behind the more work you will find in front of you as the release nears. Such neglected activities may even become a showstopper for the release. Conformity to regulatory requirements is not an afterthought activity.

12.1.2 Development Life Cycle in a Regulated Environment

Medical-device regulations do not force any specific development life cycle on you. You can meet the requirements using, for example, a waterfall-based model or an agile development model. The regulations do, however, insist on the relevant evidence being available to prove you and your device are in conformity with the requirements. In the case of the EU MDR this means, first and foremost, compliance with Annex I in EU 2017/745. Similar evidence-oriented requirements exist in other regulatory contexts. ISO 13485, therefore, requires that the manufacturer understands these requirements in their target markets and appreciates their role in appropriately defining the D&D inputs for the development project.

The regulations also expect you to have a good understanding of the medical domain your software will be used in. Manufacturers usually remember to involve a clinical expert familiar with the domain when developing the Clinical Evaluation Report (CER), but particularly smaller manufacturers may forget to involve the same expertise during the requirements definition phase. Occasionally, too, we get the question of can a manufacturer use a clinician they are familiar with to work on a product that is targeting a very different area of medicine. In practice, this may be problematic, and not an ideal choice as the goal is not to have a doctor sign off on your analysis, but rather to understand the use of your device as well as possible. In any case, involving a clinician early on in the design will in all likelihood save you from discovering some critical issues late in the development.

Occasionally, also the expected velocity of development, as well as the notion of fixing any shortcomings in a later sprint, may become an issue for regulatory compliance. It may, for example, happen that the development of the next impending maintenance update always consumes all of the manufacturer's available resources and does not leave much room for keeping up with the regulatory needs around the current version.

In adapting your agile development life cycle to appropriately address the regulated context you may want to consider the following questions.

- What is it that I value in agile development? Is it the velocity alone, or do I feel that agile is the best way to control the development process?
- Are there requirements from the standards which I feel slow down my agile model? Do I understand these requirements?

- What evidence does the management of the agile life cycle itself give me? Which pieces of this evidence are valuable in the regulatory context, and how?
- What deliverables are expected by the standards and my applicable regulations?
- Do I know where each of the deliverables fit in my life-cycle model? Do I know when each of these is intended to be ready? Do I know the point after which these deliverables are unlikely to need revision?
- How do processes such as usability, risk management, and change control affect my life-cycle model? At which points of the development work do these fit in, or where is their impact the greatest?
- Am I happy with my control over both parallel and serial work?
- How do I monitor external requirements and changes in them? For example, regulations and their guidance will evolve over time, perhaps even over the development phase of the software.

Answers to the above questions, and to the follow-up questions you will run up against in going through the above list, will be your way to figuring out if and how you should adopt an agile software-development model from a regulatory point of view. The more detailed you can get in the questions the easier you will find it to fashion the required standard operating procedures for your organization – and the better off you will be in meeting your regulatory expectations.

In practice, no life-cycle model available today will fill the regulatory requirements for you. You will, in all likelihood, need to connect the dots yourself in the way that best meets the needs of your organization and your products. Every model has its pros and cons, there's no use moping about the issues you will face, but it is your responsibility to get it all done in a practical, acceptable way. The discussion in this section was not intended to say that you are lazy if you can't handle the waterfall, or that you are stuck in the 1950s if you don't want to embrace the agile. Our aim here was to introduce you to some of the topmost points of lamentation and exuberance with either model. There are also more life-cycle models introduced in literature than we covered here. We only chose two of the most prominent examples at the opposite ends of the spectrum in terms of longsightedness and flat-out clairvoyance.

At the end of the day, the three most important questions to answer in shaping your development model, whether you start from a waterfall or a washing machine, are as follows.

- Does the model achieve what I want and get me the outcomes I expect?
- Does the model produce and maintain all the regulatory evidence I need in my chosen market areas?
- Is the model cost-effective and reliable?

By analyzing the above questions during the development of your life-cycle model you will stand the best chance of arriving at a model that not only meets the regulatory expectations but does so reliably in the long term. There are many more questions you will have to consider during setting up, running, and later updating your development model, but many of them should be traceable back to these fundamental questions.

12.2 IEC 82304-1 on Health Software

> *Films often get remakes after enough time has elapsed. Julia Ormond stepped into the shoes of Audrey Hepburn to portray Sabrina, Robert De Niro hung on to the bottom of a moving car to follow Robert Mitchum to* Cape Fear, *and Timothée Chalamet wiped away any traces of earlier versions of* Dune. *In all these cases, someone somewhere decided that it would be a good time to do a bit of a do-over. In some respects, for IEC 62304 that remake is IEC 82304-1. The new standard does not replace the old, but it does investigate some of the points of criticism levied against its predecessor – even if it appears to change the forum from the silver screen to YouTube in the process.*

The IEC 82304-1 standard on product safety requirements for health software is technically beyond our scope here in this book, but as the standard is routinely considered as an extension to IEC 62304 it is worth discussing here in some detail. As a separate document it perhaps also demonstrates some of the shortcomings identified in IEC 62304 during its prolonged stay at the maintenance dock. It should, however, be noted that IEC 82304-1 has been written with health software in mind, and this does not necessarily imply the same deep end of the pool as medical-device software – depending on who you ask health software may be the whole pool in general or just the kids' end of the pool. The standard is intended to be compatible with IEC 62304 through and through.

In recent years software has forged a definite beachhead into the monitoring and diagnosis of patient health. This foothold is only being enlarged with each new feature and new device released into the global marketplace – so much so that it is fair to say software has an increasingly important role to play in how we detect and treat diseases today. The presence of software also transcends the traditional boundaries of space as the automated analysis of vital parameters and other forms of health data may, for example, take place at the bedside, at an on-premises hospital hub, at an off-premises central analysis hub, or even in a distributed fashion in the cloud. The decision on the treatment of the patient is still made by the healthcare professional, but the information they use may be prepared by a complex network of software services. In addition to the analysis taking place in a distributed fashion the data sources used may themselves be distributed across different geographies and systems (e.g., electronic health records, laboratory systems, imaging systems, and patient monitoring systems). Furthermore, the platforms and operating systems such software runs on will be developed by third parties and often for some general purpose beyond the intended use of the medical software.

To address this modern-day diversity, IEC has published the IEC 82304-1 standard titled *Health Software — Part 1: General Requirements for Product Safety*. In films any time you see a movie labeled as "part one" you know that it's not going to end satisfactorily until a few sequels later. The standard, though, is a fairly satisfactory extension of what has come before. This standard is intended to build on the established IEC 62304 standard and set requirements for stand-alone software that is intended to run on general computation platforms. Appendix A of the standard lists examples of the types of software it can be used for, but in practice all types of software – including that complying with Article 2 of the EU MDR – can adopt the standard on top of IEC 62304. However, medical electrical equipment, IVD devices, and implantable devices are primarily left for other standards to cover.

IEC 82304-1 is widely recognized in the medical-device world as a helpful new kid on the block – someone who will help carry your groceries back home, and even spot some key items you forgot to buy, but it will not speak to the healthiness of the meal you will make out of the ingredients. In other words, the standard is a helpful addition to refer to when trying to meet the requirements of the standards and the regulations, but it doesn't attempt to offer anything close to a standalone solution.

The standard is recognized in many jurisdictions around the globe (e.g., US FDA, Australian TGA) and routinely addressed in conformity audits in one capacity or another. The standard does contain valuable elements that would otherwise be missing from the overall framework set up by ISO 13485 and IEC 62304. The thinking offered by the standard and the itemized lists of aspects it points out in relation to, for example, identifying product requirements (including those regarding platforms, security, usability, and accompanying documentation) and conducting product validation, each serve to give added structure to the development of medical devices. The thinking it offers is, however, best viewed as guidance, not necessarily as requirements to hit during your D&D activities. Adopting the standard may help you to consider all the poignant points of view as inputs for your D&D activities, and thus elevate the level of your design activities.

The standard adds to the story told by IEC 62304 and ISO 13485 so far in terms of the following five topics:

- Product requirements
- Software development process
- Validation
- Product documentation
- Post-market activities

Let's look at each of these briefly in the following.

12.2.1 Product Requirements

The determination of product requirements now includes identifying risks related to not just safety but also information security. The manufacturer must also consider, as is applicable, those situations where their product is configured or has interfaces with other products. Full ISO 14971 conformity is not required here, but the manufacturer must determine the need for risk controls (mitigations) for unacceptable risks. This risk assessment then serves as an input to product requirements. IEC 82304-1 does not expect the manufacturer to be able to adhere to ISO 14971 for each constituent part of the software (e.g., proprietary components, non-healthcare components, and legacy software), but they are expected to perform risk control around those components where risks are found to be unacceptable.

The standard also goes into added detail on use requirement and system requirements, and how the two are to be verified and updated (although here no real new requirements are introduced). Here you are instructed to consider the following new items for use requirements:

- Interfaces and user interfaces (e.g., What is to be achieved via the interface? Is the user a human or another system?)
- Immunity from/susceptibility to interference from other software running in the same environment (e.g., concurrent memory use, EMC/ESD interference)
- Privacy and security issues (e.g., authentication, malicious use)
- Accompanying documents (e.g., IFU)
- Software updates (e.g., upgrades, rollbacks, patches, distribution, deletion, and retention of data)

In terms of system requirements, the standard points out the following potentially new items:

- Technical user interface properties (e.g., display color, character size)
- Operating platform (e.g., operating system, software libraries)
- Inter-operability
- Localization and language support
- Risk controls implemented in the product (at the system level)
- Detection and handling of security compromises
- Protection of essential functions
- Product configuration, retention, and recovery

Glancing through the above abbreviated lists it becomes evident that the IEC 82304-1 standard broadens the field of vision set up by the other two standards, ISO 13485 and IEC 62304, from safety to security in quite distinct ways. The consideration of the underlying platform is now more nuanced than just some piece of hardware the manufacturer is expected to have full control over as we are looking at general-purpose computing platforms. They should, however, address the acceptable parameters of those platforms and their own risk controls for working on top of those platforms. Despite the throning of security on a level next to safety, we should not lose focus of the user and the patient as the origin of our coordinate system for risk management – IEC 62304 expects all risks to be measured in relation to the patient and IEC 82304-1 says nothing to change this. The Implication is that a business risk related to security is not necessarily in scope, but a security risk with a harmful impact to the patient or the user definitely is. News stories from the past years provide us with ample examples of security vulnerabilities having led to leaked health records and compromised patient privacy.

12.2.2 Software Development Process

The IEC 82304-1 standard adopts IEC 62304 as the basis of all its requirements. In particular, it states that Clauses 4.2, 4.3, and 5 through 9 of IEC 62304 apply in addition to its own requirements. This constitutes the bulk of IEC 62304, although technically the clauses on the QMS (Clause 4.1) and legacy software (Clause 4.4) are omitted. IEC 82304-1 further states that the system requirements identified are to be used as the primary input for the life-cycle process of the software product.

12.2.3 Validation

IEC 82304-1 speaks of validation in far more detail than either of the two principal standards discussed in this book. Here it sets requirements for how the validation is performed as well as what an individual validation plan and a report should achieve.

The plan must address all user requirements, the validation scope and activities (incl. activity feasibility and criticality), and validation methods (incl. inputs and acceptance criteria). It must also address the systems and services needed in the validation (e.g., operating environments where the software is to run), the qualification of personnel performing the validation, and the independence of the validation team from the design team. The plan must be established before validation is performed, or validation readiness can be said to have been reached.

Before commencing with the validation, the validation team must be sufficiently qualified, and the relevant IEC 62304 Clause 5 software-development phases for the parts of the software being validated are to be complete. The former stipulation is what ISO 13485 would also say on the matter, but the latter stipulation is exciting as it opens the door to piecewise validation of software. Remember that ISO 13485 is adamant that a representative sample of the final product is used for validation, but here IEC 82304-1 appears content to validate parts of the eventual product. In practice, you will want to meet both expectations: if you validate parts of the whole, you must be able to comfortably argue that also the validation of the whole is satisfactorily met before product release. In practice, you will want to carefully plan the interplay between your various verification and review activities and explain how these are planned to be useful in your overall validation activities – especially if you perform some validation during these earlier development activities. The analysis of the whole body of work can be made in the final validation report, but it is beneficial to know how the various activities slot together from the start and to make sure your software-development activities are complete enough to allow the corresponding validation activities.

The validation itself must take place as planned and in the intended operational environments. Any deviations from the plan must be justified in the report, and any anomalies discovered are to be handled via the problem-resolution process discussed previously (see Section 9). Something which is occasionally forgotten in today's agile development environment is that it is also necessary to assess the impact of the modification on the other design choices already made, and possibly launch further change-order processes to address these effects (see Section 8.2). The validation may then need to be repeated in whole or in part if the software is modified.

In terms of reporting on the validation, IEC 82304-1 requires that the validation team develops the report. This could be understood simply as a report having to be developed, but the requirement is oddly specific in requiring the validation team to be the ones developing it. In other words, no stunt performers are accepted here. The report must record the validation conditions and activities, any deviations (incl. justification) or anomalies, and the validation team (incl. name, affiliation, and function). It must then provide a summary of the results, all the required evidence (i.e., the traceability of results back to user requirements, the fulfillment of the user requirements, and the acceptability of the residual risk), and a conclusion that validation is now satisfactorily completed for the intended use based on the user requirements.

Also, note that the standard discusses the need to re-validate the software after changes in its clauses on post-market activities (see Section 12.2.5).

12.2.4 Product Documentation

As IEC 82304-1 expects your software to live a half-nomadic existence jumping from one platform to the next, it quite rationally also has expectations on the documentation that will travel along with it for the users to inspect. You can think of this as the IEC 82304-1 passport for your product containing both the necessary identification information for it and the key documentation for its use.

In terms of identification, your software must be identifiable by your name or trademark, product name or type reference, and a unique version identifier (e.g., version number, manufacturing date, UDI). The user must have access to this information when using the software (e.g., via an "about" box, on the opening screen).

In terms of the provided information, the standard expects there to be both instructions written on a user-friendly level and a more technical description of the product. The set of documentation provided must describe any requirements placed on the user (e.g., training, skills, and knowledge), the use environment or location, and the hardware or software to run the software. This is all to be based on the intended use of the device. The documentation must be user-friendly, or in other words match the user's education, training, and possible special considerations. The documentation may thus benefit from, or be required to undergo, usability-engineering assessment (see Section 12.5).

The user-friendly instructions probably take the form of the now familiar Instruction for Use (IFU). Here IEC 82304-1 expects all the topics relevant to the installation and use of the software to be covered including the following:

- The intended use, limitations for use, and known technical issues
- Installation and platform requirements
- Dependencies, interfaces, and configuration options
- Restrictions on IT network usage
- Starting up and shutting down the software
- Introduction and essential features
- Operating instructions and explanation of symbols, signals, warnings, notices, and error messages
- Safe decommissioning

Note that the warnings and notices included for safety or security reasons must be instructed adequately in the IFU – what is thought adequate is determined via the manufacturer's usability-engineering process while paying special attention to those warnings perhaps used to mitigate risks to an acceptable level. IEC 82304-1 Clauses 7.2.2.1 to 7.2.2.10 give the full list of the individual items to cover here.

The technical description, on the other hand, is to contain the following.

- All essential information (curiously called data by the standard) to safely transport, store, and operate the software. This includes any necessary information on environmental transport/storage conditions for media,

and any conditions, measures, restrictions, or special requirements necessary for installation.
- Information on supported platforms and hardware/software system requirements, and how to cope with hardware/software platform changes.
- Information on configurable security options (incl. operational options), and how to select platform settings to promote security.
- Information on required maintenance.
- Information on the effects of a security failure and its impact on patient care, data, and clinical workflow.

In addition, the description is to contain or reference all the characteristics of the software (incl. ranges, accuracy, and precision of the displayed values), which may be a tall order. Clause 7.2.3.1 gives the full list of the individual items to cover here but does so in a lamentably confused presentation.

The discussion of configurable security options grabs the attention from the above list. This may, for example, include selected network ports, password options, and logging levels. Prescribing strict rules to promote high security is a noble goal, but if this leads to poor usability, especially during installation, the users may act loose when being faced with difficulties in getting the software to work at all. Nevertheless, you should not leave doors open if you don't want visitors through those doors.

In the case of software intended to connect to an IT network not controlled by the manufacturer, additional expectations are placed on the technical description. Firstly, the provided information must now also cover descriptions of the intended information flow between the software and any other software or the IT network (incl. characteristics, configuration, technical/security specifications, and malware protection). Secondly, any hazardous situations arising from network failures are now to be listed and the user of the software is to be cautioned that further unidentified risks may arise to the user (also patient and other third party) when software is used on a network. These risks should be appropriately managed, and it should be remembered that any changes to the IT network (e.g., changed configuration, and hardware/software updates or upgrades) may give rise to new risks.

Finally, similar to the identification of your software, the accompanying documentation is expected to reference the manufacturer, the manufacturer's contact information (incl. website address), product identification, and document version. Don't interpret this as a need to always stamp this information on each piece of the documentation, but to include the information somewhere practical within that body of documentation.

12.2.5 Post-Market Activities

The IEC 82304-1 standard intends to address the entire life cycle of software all the way up to maintenance, decommissioning, and disposal, as stated in Clause 1. Yet, in terms of assessing the compliance of post-market activities, it draws the line at aspects related to product D&D (Clause 8.1). The intention here may be to limit the scope of information collection and analysis needs to some practical sphere around D&D, but the wording is cumbersome. Given that we know sensitive data has occasionally found its way into scrapyards when some disk drives or memory media have not been wiped appropriately, we should also concern ourselves with learning from such incidents to better design our devices and instruct their end of life. Given that users are not naturally giddy to analyze and install each new version update put out by maintenance, we should think carefully about how we want to instruct critical software maintenance. The standard does go a long way to addressing all these aspects in its other clauses, but when it comes to discussing them in the context of post-market activities it appears a little coy here.

On maintenance, the standard invokes IEC 62304 with the disclaimers it made earlier in its Clause 5 (e.g., IEC 62304-compatible development model, ISO 14971-lite risk management). If maintenance is required, also re-validation must take place as is relevant to the modification and the supported platforms, and the accompanying documentation may need to be updated. The standard also calls for the validation plan to be updated accordingly after re-validation, but whether that results in changes to your plan will depend on your plan – only update plans when your plans change, not to update timestamps or to just track their execution. Remember, too, that the detected safety and security issues may also trigger regulatory reporting needs and actions in revising the conformity assessment.

On product-related communication, the standard expects to see the manufacturer take steps to inform users about a) security vulnerabilities, b) changes in relevant regulatory requirements, c) available new versions of the product, and d) a range of other information such as new features, fixed issues, safety or security developments, and updates to software identification or accompanying documentation. The decision by the user to install an update should be based on the safety and security of the new version.

Appropriately, if the new update has a positive impact on safety and security, the manufacturer may advise the users to replace their older versions with it. The wording of the clause here is less than ideal, but it appears to discuss both communications accompanying a new release and the communication taking place to, for example, enable the discovery of that release by its intended users.

Finally, the standard addresses the skirted decommissioning and disposal of products. Here the manufacturer is expected to consider applicable use requirements and provide a safety function for disposing of the product. This necessarily includes any measures needed to safeguard personal health data and other security and privacy concerns.

12.3 Segregation in Software Architecture

> *When Riggs went berserk in "Lethal Weapon", he had Murtagh to talk him back from the final ledge. When the adults in "Jurassic Park" were at a loss to know how to proceed, the girl pulled in a trump card of knowing Unix. And when the reigning Phantom kicked the dust, the next generation was again there to pick up the slack just as promised by the old jungle saying. None of this would have been possible if all the actors would have been handcuffed together all along.*

The term software architecture (see Section 5.3) refers to the composition and internal structure of software, including how the various subcomponents communicate with each other. The structure may, for example, be defined along a functional axis (i.e., what components are involved in the realization of what function or feature of the software) or an ontological axis (i.e., what components are used by which other components). Occasionally the difference between these two axes may become blurred, but what matters is that the developed architecture satisfies René Descartes to the extent practical for realizing both the individual components and the whole of the software.

As discussed in Section 1.5, in the eyes of IEC 62304 everything in the software from units to the whole of the software system is fundamentally a software item. Each item has its own part to play on the whole, but in addition to their correct function, we must also address their possible negative effects on the whole. What happens if that item malfunctions? Can it cause harm to the users, patients, and environments around the medical device? This is where the concept of segregation becomes extremely important. In other words, this is where we realize there's a Fredo in the architectural design and they are courting disaster so we must cut them out of the family tree like Michael did in *Godfather II*. We already briefly discussed the bulkheads onboard the Titanic and the role external risk controls can play in addressing risks in Section 4.3, but let's dive a little deeper here.

Traditionally segregation has perhaps been thought of as a property of physical components. Redundancy has been built into systems to ensure that, for example, critical life support processes in hospitals or naturally hostile environments such as space do not accidentally end on a glitch. Redundancy may have meant adding parallel components such as multiple processors to a device or perhaps even just adding an uninterrupted power supply to take over should a mains connection fail. Today, software is responsible for many critical actions in medical devices, and a flaw that goes unchecked in one component may cause serious harm depending on the type of device. It is a common mistake to think that software errors can only be caught by hardware controls or instructions to the user. Not so. Software can, and in some cases should, watch over other software. This is why segregation between software systems and possibly software items is such an important topic today.

Standards such as IEC 60601-1 and IEC 62304 do speak of segregation. The primary setting for this discussion is risk management, where the role of reliable segregation is understandably heightened. When the risks regarding one software item are mitigated (controlled) via expectations placed on another software item it becomes extremely important for the second item to be as immune as possible to the failure of the first. According to IEC 62304 Clause 5.3.5, the manufacturer must identify such segregation needs regarding class C software items, and also ensure the effectiveness of the segregation. The standard gives us a simple, if non-binding and non-exhaustive, example of running the two software items on separate processors that do not share any resources as a reference to effective segregation. Figure 12.1 illustrates a possible breakdown of this solution where the processes from each processor have access to the same memory space.

The example given by IEC 62304 is not exhaustive and it is not much more satisfying than saying that two items separated by lightyears of space between them are not likely to affect each other unless you believe in the butterfly effect. Luckily, IEC 60101-1 is more forthcoming. Clause 14.8 of that

278 ■ *Medical-Grade Software Development*

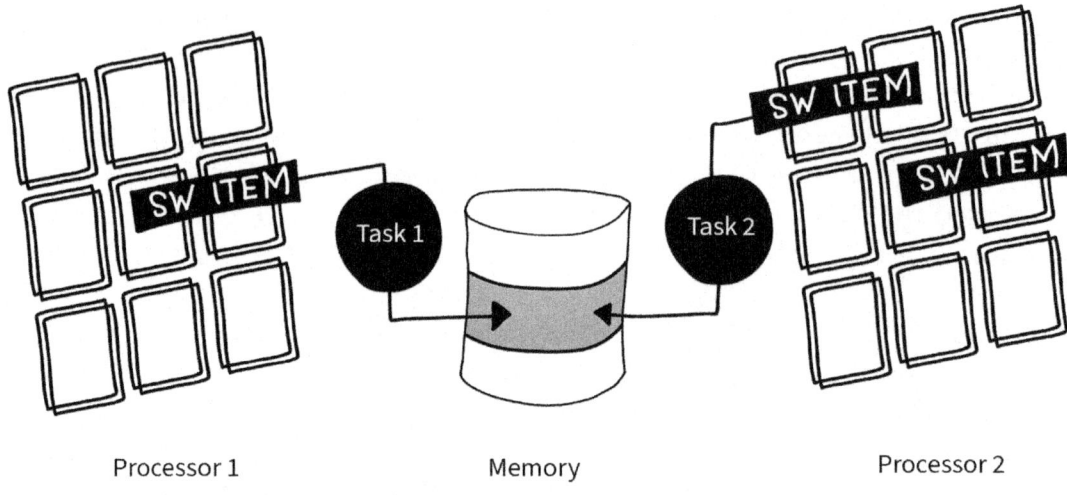

Figure 12.1 Segregation breakdown due to memory access.

standard lists the following design options for mitigating risks in the device architecture:

- Using high-integrity components (e.g., components with high mechanical durability)
- Employing fail-safe functions
- Building in redundancy
- Allowing for diversity
- Partitioning functionality
- Incorporating defensive design (i.e., countering harms via pre-emptive design choices such as limiting electric currents and the physical movement of actuators)

From the above list, it is apparent that IEC 60601-1 is considering physical medical devices, not software, but the thinking it offers is valuable in both contexts. For software, the list could, for example, refer to using validated software tools and controlled SOUP components and building in fail-safe functions, redundancies, and defensive design considerations. The take-home message here is that it is entirely possible to improve the resilience of software and reduce the risks in its use via designing for safety in case someone still had their doubts on the matter. How effective, necessary, or sufficient those methods are will depend on the type of medical device.

The IEC 60601-1 standard goes on to suggest that the following are considered when designing the architecture: a) allocation of risk-control measures to subsystems and components such as sensors, actuators, and interfaces, b) failure modes, c) common cause failures, d) systematic failures, e) test interval duration and diagnostic coverage, f) maintainability, and g) protection from reasonably foreseeable misuse.

Based on the above, good segregation is not limited to physical measures (e.g., redundant processors and memory partitions). Despite the scarcity of examples in IEC 62304, segregation measures include all appropriate mechanisms that act to prevent one piece of software from negatively affecting another. The adequacy of this segregation may then be determined based on the associated risks, risk-management activities, and risk documentation. Even when relying on memory partition as a segregation method the manufacturer must still consider all risks related to that function, including the allocation of memory. An example here might be that a Task 1 running on CPU1 is not able to access the memory space used by Task 2 on CPU2 but it is able to somehow affect the memory allocation criteria for CPU2. The example here is not meant as fearmongering, only as a friendly reminder that magic bullets do not exist, and it depends on the nature of your application and how tight you should wind the tinfoil around your head.

The discussion in this section has brought up aspects to consider when specifying segregation in your software. The lens we have used here is risk management where segregation is naturally of the highest importance. We don't really care if Riggs and Murtagh share a slice of germ-laced pizza every evening after work, or if Phantom number 35 had a happy childhood, but we do care if the segregation we rely on is not really there in reality. The point here is both that solid segregation may be necessary for reliable risk control to take place, and that addressing any potentially significant leaks in the segregation of components and subsystems is a part of risk management. The place to analyze and set up controls for segregation is in developing the architectural design for your software, which also has to be in sync with the requirements defined for it.

In the world of physical medical devices, segregation may have been mostly about electric shock and EMC testing. In simple software, this may be expanded to segregation between software items of varying degrees. In complex, network-connected, and perhaps constantly evolving devices, though, risks may also be introduced via changing or misbehaving datasets, not just misbehaving code. When a decision support system, for example, bases its assessment on available reference datasets, and these change over time, the analysis offered by the software may change as well. If the dataset is no one thing that changes through careful curation, but instead the reference dataset or learning dataset consists of multiple information sources pooled together the likelihood of issues propagating may also increase. Data quality criteria may be part of the solution to working with such shifting foundations, but so too may many of the segregation design considerations introduced above for hardware devices. On other occasions, addressing cybersecurity issues in some appropriate way may be key (see Section 12.6). In the long run, we will also need a good standard for medical-device data governance and use to augment the thinking offered by ISO 13485 and IEC 62304.

12.4 Risk Management Influenced by FMEA

> *Asteroids and natural disasters are probably more of a concern for FEMA than for FMEA, but for analyzing risks in man-made constructs the latter has a very strong following. FMEA is no stranger in an ISO 14971 setting, either, but in* Tropic Thunder *terms it is the Tom Cruise of the show, not Ben Stiller or Robert Downey, Jr.*

Failure modes and effects analysis (FMEA; see IEC 60812) is a method developed in the 1950s to identify potential failure points in systems consisting of multiple components and subsystems. The aim is to identify the failure modes of each component and analyze their effect on the whole. FMEA is a popular bottom-up method that analyzes each constituent part at a time, and once a potential failure mode is identified, looks at the hazards and harms that may arise from it, as well as the initiating cause behind the failure. Each sequence is then evaluated based on the severity and probability of the harm.

In medical devices, FMEA is often used just as it is used in other domains of engineering, although more than one flavor of FMEA exists in each of these domains. It is, however, crucial to note that FMEA does not necessarily cover all the aspects of risk management expected in medical devices. As discussed in Section 7.5, risk management in this context must consider at least the following types of risks:

- Risks associated with software failures
- Risks associated with the correct, intended use of software
- Risks associated with foreseeable misuse of the software

Of the above types of risks, FMEA is most readily applicable to software failures, of course, but it may miss important risks falling into the other two categories. In the context of medical devices, the base standard expected of all risk-management activities is ISO 14971, which can be leveraged for all three categories. Techniques such as FMEA may be used to help in this overall analysis but are usually not comprehensive enough to tackle the whole sphere of risks by themselves. It is up to the manufacturer to set up their risk-management process so that it meets the combined expectations of the standards discussed in this book (mainly ISO 14971, ISO 13485, and IEC 62304) and produces all the relevant information expected by the regulations. In this, FMEA is not the magic bullet solution some appear to expect. It is, however, a good tool to have in the complete toolkit.

Considering that an FMEA must meet the expectations of ISO 14971 to pass muster in the medical-device context, we should consider what elements the two have in common. In practice only the starting point may be different between the two: whereas ISO 14971 anticipates risks from a wide field of sources, FMEA has its sights on the parts of the device. Both speak of hazards, causes (in FMEA: initiating cause), harms, probabilities, severities, and a score calculated for each risk (risk priority number or RPN for FMEA; see discussion below). Thus, a practical way to combine the two worlds is to allow for multiple different types of risks, one of which is component failure. Figure 12.2 illustrates this. In the figure, failure modes identified for components are one of the sources of identified risks, just like the intended use, usability, foreseeable misuse, and cybersecurity of the device.

Note that the figure here omits requirements for the sake of clarity. The intended use shown in the diagram should in many ways act as the ancestral root of all information on the developed device, including how it is

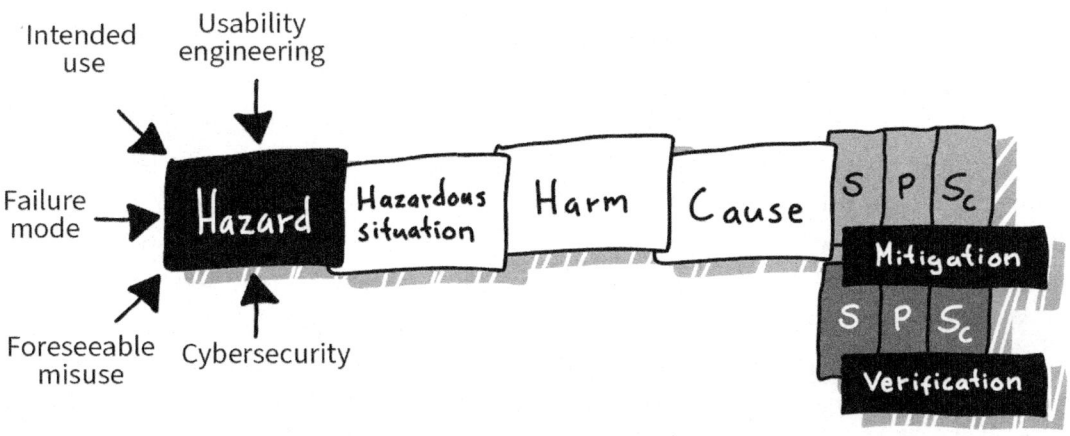

Figure 12.2 **Hazards as a gateway to risk management.**

ultimately to be qualified and classified as a medical device under the applicable regulations. The intended use would then spell out requirements for the device, which in turn would lead to identified failure modes and which could then contribute to hazards. At the same time, the intended use would advise the preparation of a use specification (See Section 12.5), which in turn could lead to the identification of new hazards. Thus, the figure here is a simplification drawn up to illustrate the general role these types of information play in risk management.

The risk score (S_c), as expected by ISO 14971, is often a simple multiplication of the severity (S) and the probability (P) of each risk. Both severity and probability may be expressed as percentages (0.0–1.0 or 0–100%) or on a chosen practical scale (e.g., 1–5). The score is then a multiplication of these leading to a combined classification on a scale from 0 to 1 (0 to 100%) or the square of the scales used originally (e.g., 1–25). The risk priority number (RPN) introduced by FMEA is similar but not the same as this simple score. The chief motivation behind RPN is to differentiate between some risk that is severe but rare (e.g., severity 5 and probability 1) and a risk that is negligible in effect but likely (e.g., severity 1 and probability 5). Plotted in a matrix of severity and probability these two risks would occupy opposite corners of the matrix, but as numbers they would look the same. To tackle any confusion, the RPN scheme assigns user-defined scaling factors to either a) the severity classification alone or b) both the severity and the probability classifications. This way the RPN scheme separates such risks at a glance. For example, by assigning a multiplier of 5–25 to the severity classification, the combined score for a severe but rare risk becomes 125 (25*5*1) and the score for a negligible but likely risk becomes 25 (5*1*5). For comparison, the RPN score for a risk that is both severe and likely would be 625 (25*5*5). Figure 12.3 illustrates the calculations involved in the determination of the RPN score.

Similarly, we could also assign arbitrary multipliers to the probability classification if this provided some practical separation for us in risk management. Also, remember what we discussed in Section 4.3 about software always failing, and if you decided to extend the P1=100% assertion beyond simply assigning the software safety classification act accordingly here.

In practice, FMEA may capture much more than just a third of the medical-device risks expected to be managed. FMEA may also be the best, most practical, and best-known method for tackling risks related to software failures. In such cases, the gap between FMEA and ISO 14971 risk management may also be assessed via a summative risk-management report that addresses

Seeing into the Future ■ 283

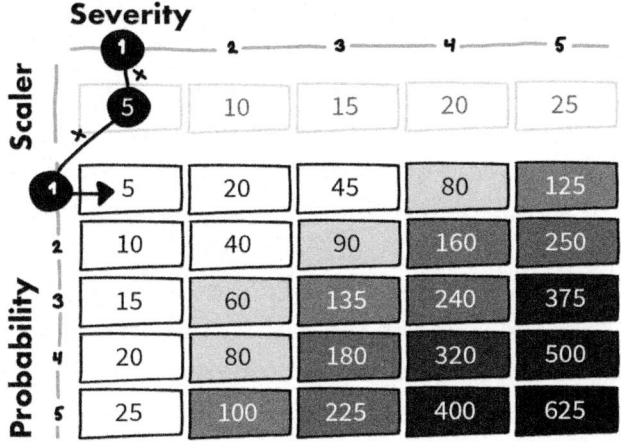

Figure 12.3 RPN risk scores scaled to emphasize the role of severity.

any FMEA shortcomings when viewed from an ISO 14971 perspective. The report must then demonstrate that the overall risk management (incl. residual risks and the overall residual risk) is at an acceptable level.

In assessing how, and to what extent, FMEA can be used to meet medical-device risk management needs it is important to consider the types of devices manufactured, their intended uses, their features, and the technologies they incorporate when attempting to identify the risks involved in their use. Component failures are a part of the picture, but as illustrated by the list above this may not be the whole picture. Herein lies the principal shortcoming of FMEA in the medical-device domain: how can we reliably capture all the relevant risks and harms that may befall the patient, the user, and the environment via analyzing component failures? And if we are too tuned into component failures, will we be working overtime on risks that are not actually medical-device risks in that no harm can come to people? Both of these issues can, of course, be solved with FMEA, but neither may be a simple task when the device and the context of its use are as complex as is to be expected in the clinical setting.

One potential hurdle to cross when adopting FMEA is, therefore, the need to align activities and outcomes with the ISO 14971 standard which sets the overall framework of expectation in medical devices. At the same time, more than one flavor of FMEA exists which means the manufacturer must choose from multiple analysis matrices and perhaps still modify these for their own use. This can all be done, and it may lead to the best solution for risk management, but FMEA is not the be-all and end-all of medical-device risk management. For more discussion of FMEA see IEC 60812.

As popular as FMEA is, it is only one possible method to use. Another popular option is fault tree analysis (FTA; see IEC 61025). This is a top-down method for identifying risks resulting in harm to patients, users, bystanders, and the environment. The analysis is performed via so-called top events, which are an abstraction of faults that can then be linked with hazards, hazardous situations, and harms. For more discussion of FMEA see IEC 61025.

Both FMEA and FTA are standardized methods and are described in detail in their associated standards and guidance. Both can be leveraged in the context of medical devices, but it is worth remembering that in the end whatever method is chosen for use in medical-device risk management must meet the expectations of ISO 14971 and the applicable regulations. As a friendly warning, also note that not all standards agree on the definition of risk: business risks, for example, do not generally feature in medical-device risk management unless they may cause harm to the patients and the users of the device, but such risks may be given as the primary reason for the existence of some other standards developed on other domains.

The discussion in this section should not be interpreted as an attack on FMEA, only as a friendly warning that reaching the Rockies does not really mean you have driven the length of Route 66 no matter which end you started from. ISO 14971 gives you the yardstick you will need to use here in each and every case, but methods like FMEA and FTA may help you cross the finish line in style – just like Tom Cruise helped wrap up *Tropic Thunder* with a bang.

12.5 Usability Engineering

> *Good usability in medical devices means safety over snazziness. It is the med bay Noomi Rapace climbs onto to single-handedly perform an abortion of the alien lifeform, not the futuristic floating displays Tom Cruise commands like he's the conductor of a great symphony orchestra. Equating coolness and usability is an easy mistake to make, particularly when entering medical devices from another industry.*

The usability of medical software is a topic often addressed from the point of view of risk management, and rightly so. The different jurisdictions and market areas across the world may each impose requirements on the usability of medical devices. Such requirements are usually not set from the point of view of intuitivity or ease-of-use, but with a view to avoiding use errors stemming from bad design (e.g., Clause 5 in EU MDR Annex I). In medical devices the terms "usability" and "performance" both actually refer to safety, one from the outside in and the other from the inside out in reference to the device.

In medical devices, usability is primarily about understanding the interaction between the device and its user or user group. At this interface a simple erroneous action performed on the part of the user may affect the device in unwanted ways, and similarly the user's incorrect understanding of what the device is saying may misdirect subsequent actions in the care of a patient. Both errors may lead to a hazardous situation or a right-out use error. Neither error can be chalked up to mere user error, which is why the term use error is used instead and the software is also on the hook as a source of contributing factors.

A user interface thus acts in two opposite directions. It is the medium through which the device presents information to the user and at the same time the means for the user to manipulate the device. In terms of usability, it is both important to ensure that the information is presented in a clear, appropriate, and readily interpretable way, and to make sure the actions offered to the user also make sense to them before these are invoked. Clarity here means, for example, that numerical display can't be misinterpreted from an off-axis angle (as has famously not always been the case with recessed segment displays), and appropriateness in tuning the offered information to the skills, education, and mental context of the user. It is in the best interest of both the user and the device to make this interaction as predictable as possible, especially in terms of what is going to happen after each particular user input or action. The science of usability engineering – whether that takes the form of human–computer interaction, service design, or some other appropriate method – is how we formally tackle this interface and attempt to arrive at clear design requirements in the form of, e.g., use specification, UI specification, and hazardous use scenarios. Methods we can use in this work include task analysis, simulation, function analysis, contextual inquiry and observation, surveys, interviews, expert reviews, and advisory panel reviews.

The methods we might use each have different characteristics and may vary in their suitability to the different stages of a usability study. Out of all the methods, task analysis is perhaps the one best suited to modeling the way users interact with the devices. In task analysis, we investigate this interaction via the use of scenarios that attempt to capture the tasks the user wants to perform and help the manufacturer design matching user interfaces and identify the risks involved. As issues are detected that may lead to use errors these are analyzed further. In practice, the manufacturer may choose the methods that are best suited for their type of product, type of organization, skillset, resources, and past experience. It is, however, important to be able to demonstrate adequate competence in the use of the method down to the personnel-level and to maintain sufficient documentation.

Typical examples of areas where shortcomings are identified in the usability of software include the following:

- Labeling (incl. visible and audible prompts for interaction)
- Navigation (incl. drop-down menus)
- Alarms (incl. alarm fatigue)
- Presets for presumed user groups
- Configuration performed by the user or maintenance staff

Many of the above types of issues are shared by both software and hardware, and it would be wrong to assume software would be immune to any of the issues. IEC 62304 itself is adamant that all software fails (see Section 4.3), which may or may not be the case, but it certainly further heightens the need to ensure good usability, transparency, and understanding between the device and its user. These characteristics will only become more important as we adopt artificial intelligence and other technologies enabling devices to evolve while in use.

A particular subtopic for usability engineering is the use of labeling as part of the overall risk-management activities for a device. If labeling is used to implement information for a safety-type risk control, it must be analyzed as part of usability engineering. ISO 14971 Clause 7.1 and IEC 62366-1 Clause 4.1.3 both agree on the matter.

The IEC 62366-1 standard on the application of usability engineering to medical devices is a cornerstone for ensuring safety through usability. The associated technical report, IEC/TR 62366-2, provides further discussion on the standard, its terminology and requirements, and the usability-engineering methods to use in the various stages of work. IEC 62366-1 sets the following requirements for addressing medical-device usability.

- The manufacturer must have a usability engineering process in place. This covers the use of the device when used according to its accompanying documentation. The process is to be implemented by qualified personnel.

- Any risk control is performed in accordance with ISO 14971 Clause 7.1 and should consider, for example, inherent safety, protective measures, and information for safety as possible controls.
- If information for safety is used as a risk control this is verified to be perceivable, understandable, and to support the correct use of the device (incl. by the correct user groups and in the correct environment).
- The conducting of usability-engineering activities is planned, and appropriate for the stage of the device or the magnitude of the change (e.g., a new device versus a small UI update to an existing device).
- The appropriate documentation is maintained in a way that facilitates the use of the usability-engineering information in the technical file of the device and as required in the applicable jurisdictions.

When a manufacturer fires up the usability-engineering process for their device, the first step expected is the preparation of a use specification for the device (IEC 62366-1 Clause 5.1). This specification is to include the intended use, intended patient population, intended part of the body or tissue type interacted with (if any), intended user profile, intended use environment, and the operating principle of the device. Note that this information is also needed in the technical file of the device for regulatory purposes.

The specification is then used to identify a) the characteristics related to safe use of the device and any potential use errors (Clause 5.2), b) known or foreseeable hazards and hazardous situations (Clause 5.3), and c) the use scenarios involved in the hazards (Clause 5.4). Figure 12.4 illustrates the general relationship between the intended use, use specification, and the task descriptions used in task analysis. In the figure, analysis of the intended use of the device first leads to identified characteristics for the use of the device

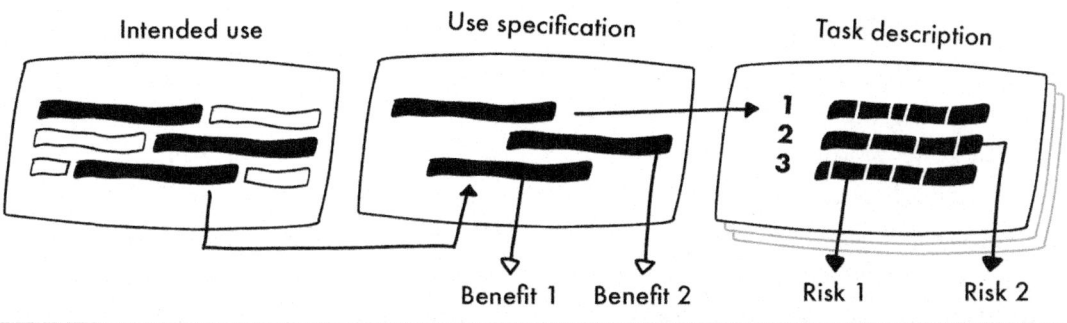

Figure 12.4 Deriving use specifications and task descriptions from the intended use.

(defined as a specification and individual tasks) and may then lead to the identification of benefits and risks for the device.

Task analysis is a good method for developing task descriptions. This analysis does not just develop a valuable picture of the user and their context, but it also facilitates the subsequent development of an appropriate user interface design and lays the foundations for a solid analysis of risks and benefits around the use of the device. Task analysis thus builds a view of what the user groups for the device are, how they interact with the device, and what features they will use. This may also include the identification of physical and social elements of the environment as well as its soundscape, if relevant.

After appropriate modeling of the use scenarios, the scenarios now earmarked for inclusion in the final summative evaluation are selected before moving into the initial formative evaluation phase (Clause 5.5). This selection may include all of the scenarios or just a subset based on some sound, documented criteria (e.g., those scenarios which can lead to a user error and harm). Next, the corresponding user interface specification and evaluation plan are developed, the user interface designed and implemented, and the initial formative evaluation performed (Clause 5.8). Finally, the ultimate summative evaluation is performed (Clause 5.9).

The two evaluations performed, the formative and the summative evaluation, may be singular activities or consist of several activities each. Formative evaluation refers to the assessment done while working on the user interface design. The objective here is to identify the strengths and weaknesses of the design, as well as any unexpected use errors. In all likelihood, this is an ongoing iterative process that takes place throughout the development phase. Summative evaluation, on the other hand, takes place after the user interface has been developed. The aim here is to create objective evidence on the safe use of the interface. This final summative evaluation must also identify any potential difficulties and near-miss situations if these may result in a use error and harm. All of the activities here must be documented, and the relevant documentation linked to the technical documentation of the device. If the summative evaluation concludes that issues exist and redesign activities are required, summative evaluation will be appropriately updated after these activities are ready.

In practice, it is possible to meet the usability-engineering expectations in one of two ways. Firstly, if the device in question is new, the user interface and the user interaction around it are assessed in accordance with the body of IEC 62366-1 (see Clauses 5.1 and 5.2). Secondly, if the device has already been on the market for quite some time, the concept of usability of unknown provenance (Clause 5.10) may be employed as described in Appendix C of IEC 62366-1. In the latter case, the preparation of the use specification is followed by a review of available post-production information, and the assessment of the hazards, hazardous situations, risk analysis, risk control verification, and benefit–risk analysis based on all of the available information.

Regardless of the chosen path to conformity with the standard, revision of the manufacturer's standard operating procedures is likely to be needed to ensure appropriate attention to usability engineering. Addressing usability will lead to specific activities in requirements specification and design, risk management, and documentation (incl. the technical documentation of a device). Usability may also add considerations for compiling the accompanying documentation for a device as this affects, for example, the installation, use, and maintenance of the device. By the same token, having a clearly instructed usability engineering process may sharpen and even streamline the design process as, for example, task analysis is used to model user interaction with the device in a more structured way. This way it may be both faster and safer to develop the correct requirements and designs.

Usability engineering is a fundamental activity with implications for several of the software development life-cycle phases discussed elsewhere in this book. The magnitude of this impact is perhaps self-evident when we consider that usability issues will factor into at least the requirements specification (Section 5.2), design (Sections 5.3 and 5.4), and risk-management (Section 7) activities performed during software development – and really should be fused into the instruction of the processes and the planning of the product development project itself. Similarly, usability engineering has clear links with post-market surveillance and change management activities as well.

Usability engineering, however, should not be seen as something extra to do. The fundamental task of usability engineering is to understand the user, how they will interact with the device, and for what purpose. As a goal, this is 100% in line with what ISO 13485 and IEC 62304 expect to see the manufacturer strive towards. Usability engineering methods can, in fact, help us model the use context better, smarter, safer, and faster than what might otherwise be possible. Methods such as task analysis may become the key solutions to understanding not only the risks but also the benefits surrounding the use of our software in the real environment. As such, these methods may give us real rigor in answering not just the risk assessment but also the benefit assessment required of us as manufacturers and making the ISO 14971 benefit–risk ratio assessment in a solid, defensible manner. Usability is not a tick-box activity, not really – not when it can offer us real insight into the way our products should work and how they are in fact used out there in the real world.

12.6 Cybersecurity

> *Johnny Mnemonic showed us a nihilistic future where cybercrime had taken us to extreme measures in protecting information. Minority Report took us to another extreme where crime was abolished by treading on the privacy and free will of people by jumping on a crime before the gun was ever picked up. Neither of these futures has yet become a reality, but the topic of cybersecurity continues to both fascinate and scare us. For medical devices the stakes are high, and the effects are possibly immediate.*

Cybersecurity is one of the hottest topics in medical devices today and perhaps only second to the use of artificial intelligence. In the short term cybersecurity will probably have a much greater impact on medical devices than artificial intelligence as it touches on just about all medical-device software whereas artificial intelligence is still in many ways more of a promise than a practice. Both are, however, shaping our environment today and will continue to do so to increasing effects in the future.

Cybersecurity for a standalone, disconnected device may be a non-issue, but when that same device is connected to a network – be it a local network or the world-wide-web – the risk of cyber-related issues becomes a distinct possibility. These in turn lead to new requirements being imposed on, for example, the design, implementation, testing, accompanying information, and maintenance of that device. Cybersecurity, much like usability engineering, is not an optional extra to the development of your medical device. It is part of the picture for developing that medical device in a way that allows you to rely on the benefits and risks assessment you have made and the value promise you make to your customers.

Risks arising out of compromised cybersecurity may be diverse and range from financial risks (e.g., damage to the brand), to privacy risks (e.g., loss of sensitive patient data), to possible patient-safety risks (e.g., hacked pacemaker software). All of these risks are serious and may lead to very measurable harm coming to both people and property, but out of all the risks, the most disconcerting ones regard situations where serious, life-threatening harm may rapidly come to patients. These are cybersecurity problems leading to the loss of some critical systems during their use, for example, a BitLocker-lock of some control software during a surgical operation, or breakdowns in life support devices at hospital ICUs. In these environments, a vulnerability may affect an obvious, essential system that has direct influence over the life or death of patients (e.g., respirators, vital signs monitors), or it may concern a less-obvious support system where breakdown may only cause annoyances or disruptions in the environment (e.g., the unavailability of some patient records), but which may end up being fatal due to the very nature of the critical environment.

If even a small cybersecurity vulnerability can be exploited for great loss, and if some cybersecurity issues may even be unfortunate accidents instead of malicious attacks (e.g., congestion and interference caused by general-purpose IoT devices in a medical environment), it becomes self-evident that cybersecurity is a very real topic to address. The goal here should be one of prioritizing risks, of course, but also preventing all cybersecurity issues to the fullest extent possible. Small fissures may become large crevasses over time, particularly when some malicious intent drives the raptors to try the cages.

12.6.1 Cybersecurity Over Product Life Cycle

In practice, addressing cybersecurity calls for actions during the design, production, and maintenance of devices – decommissioning too unless we want to see troves of sensitive information end up in landfills, which has been known to happen.

During design, all relevant sources of information including the prevalent state of the art are considered in defining the requirements for the device and designing both the device and its maintenance over the coming years. The developing software architecture and the technologies used should be analyzed from the point of view of safety and security from an early point on as the different choices made here (e.g., defensive design) set the scene for what is to come later. Techniques such as penetration testing and threat analysis may be useful here, and the basic risk-management framework instructed by ISO 14971 is a must. Clause 7 of ISO 13485 will do its part to ensure a sound design is reached, but also standards such as IEC 82304-1 (see Section 12.2) may be applicable based on the type of the device. Understanding the user, the patient, and any operating environment will be crucial here, and usability engineering may provide a part of the necessary picture (see Section 12.5). The term cybersecurity may not always leap at you from the standards and regulations written on medical devices, but you can expect it to always be a part of any consideration of risks around medical devices.

The production phase of the device should be simpler to handle than the design phase, but from here to the basic cleanliness requirements set in ISO 13485 should be interpreted as a call to ensure the software ships without malware and viruses. This may be as easy as cloning a binary or as complicated as assessing all of the new versions of the constituent SOUP components. Note also that the installation phase of a device may see drastic changes in expected settings if the configuration is allowed. Some critical security settings may be relaxed (e.g., unnecessary ports left open) just because it is easier for the installer to get it all working that way. Similarly, once something is up and running, it may be all too easy to forget about watching over it or tending to security updates later. All of this may, at least in part, be countered with the appropriate accompanying information and training provided to the user along with simple mechanisms for checking on the availability of updates.

That brings us to the next phase in the life cycle of software: maintenance. There is no discernible wear and tear involved in software, yet maintenance is just as important as for physical devices whenever that software is exposed to the elements of a network environment. In an environment that changes constantly the software itself may need to change to stay up-to-date and safe. Maintenance here refers to, first and foremost, keeping the device safe by reacting to issues quickly and providing reliable updates to the device. Providing feature updates to devices only comes second, although a manufacturer might internally view maintenance work from exactly the opposite frame of mind: as something leading to new features, not shoring up old features.

It may be tempting to approach cybersecurity as a later-stage activity performed toward the completion of a piece of software. In some ways, this is because the post-market phase is definitely implicated in cybersecurity activities, and the closer we are to that phase when we perform the final pre-release assessments the more up-to-date this work is likely to be when the product ships. It may, however, be the case that some threats could have already been thwarted in the requirements specification or architectural design phase of the software. By the same token, issues discovered near the launch may require adding new requirements, new test cases, and new verification or validation activities. If software development was a game of snakes and ladders this would definitely be a vicious snake to encounter. Instead, you should look to modularize your device and rely on already completed cybersecurity assessments available for some of those modules as the ladders to improving your responsiveness in tackling any newly detected issues.

Reducing the level of complexity to some manageable level will likely be key to maintaining cybersecurity activities on a practical level. This means modularizing your software in some smart way so that you can reuse modules and tend to their maintenance on different clock cycles as needed. This means utilizing your resources so that threat modeling for your device, for example, is not your only source of information on current cyber threats in your environment. Cybersecurity will be part of your post-market activities going forward, and like those post-market activities, it should not just be up to the original development team to keep abreast of all relevant developments. Some tasks here will be up to the engineers, but some other tasks are more akin to market intelligence activities.

Involving a sufficiently broad base of experts with appropriate backgrounds may also be important for spotting cybersecurity threats surrounding your software. Expertise in the clinical setting, usability engineering, software engineering, risk management, and cybersecurity management may all be called for at various stages during the life cycle of the device.

12.6.2 Cybersecurity Across Jurisdictions

Cybersecurity has been a hotbed of activity around the world over the past few years. A wealth of thought has been published on the topic by, for example, the FDA in the US, the Medical Device Coordination Group (MDCG), and the joint organization of European notified bodies, Team-NB. All of this published information, as well as the underlying regulations themselves, will be of interest to you as you look to address cybersecurity in the jurisdictions. In Europe, the new Cybersecurity Act, too, will demand attention on this front.

The international standard on risk management for medical devices, ISO 14971, has already seen the elevation of data and systems security to a level on par with the human-centric safety focus it had previously. National guidance and proposed rules are following suit and beginning to set direct requirements for medical-device manufacturers. It should also be noted that regulations beyond those directly discussing medical devices and cybersecurity issues may add related requirements. A case in point here is given by the EU GDPR and the US HIPAA legislation.

The good news here is that international standards are seen as facilitators of global harmonization. A position paper on cybersecurity by Team-NB, for example, refers to IEC TR 60601-4-5 for determining product requirements and IEC 81001-5-1 for both cybersecurity-related product requirements and life-cycle activities. The former document discusses both hardware and software to provide further guidance on the design of medical devices for use in medical IT networks while the latter standard is fast becoming a reference for addressing cybersecurity. Let's take a closer look at the latter standard in the following.

12.6.3 IEC 81001-5-1

The IEC 81001-5-1 standard on security activities over the life cycle of health software is currently gathering momentum as a reference for not just health devices but also medical devices. The 50-page document sets requirements for both the development and maintenance of health software but does not speak to the contents of the accompanying documentation for a device. The standard does expect all activity to take place under a quality management system implemented according to ISO 13485 (or equivalent).

IEC 81001-5-1 has been organized so that it matches with the clause structure of IEC 62304 as closely as possible, but it does not conform to the same use of the software safety classification. The requirements given by IEC 81001-5-1 are said to apply to all classes of software.

The impact of IEC 81001-5-1 on the IEC 62304 development of software depends on the type of product in question and the extent to which you have already addressed cybersecurity in your software development processes. In general, the standard does not revolutionize the approach to software development, but it does act as a good reference calling attention to sources of cybersecurity risks, and the appropriate way to handle these.

For the software-development process (see Section 5), the standard performs a comprehensive sweep touching on the determination and review of requirements, the development of the software design, the performed testing (incl. threat mitigation testing, vulnerability testing, and penetration testing), release, and decommissioning. Interestingly the standard also specifically speaks of managing conflicts between testers and developers, ensuring file integrity, and controlling private keys, among other security-related topics.

For the software-maintenance process (see Section 6), the standard describes its own expectations for the process, including the timely delivery of security updates to the software. On the agenda are also, for example, monitoring public incident reports, update verification, update documentation, and update delivery. Note that in terms of documentation, the instruction here is brief and in line with what IEC 82304-1 expects.

For the risk-management process (see Section 7), the standard naturally brings a list of new risks to the table. The effects here may be profound if cybersecurity was not already a part of the picture for risk-management activities based on ISO 14971. Here the standard attempts to help you identify, estimate, control, and monitor the security-related risks surrounding your software. The discussion is written with a view to pointing out relevant aspects, but it does not go into too much detail on how to address these in a systematic manner. Out of all the sections of the standard this may be the clause with the biggest impact on your activities due to the omnipresence and importance of risk management in medical-device development, particularly for a network-connected software-based device.

For the configuration-management process (see Section 8), the discussion in the standard is light. The requirements set here boil down to the ability to produce a list of external components used to the extent these could become susceptible to vulnerabilities. The discussion here is only five lines long and in line with the expectations already given by other standards.

Finally, for the problem-resolution process, the standard wants to see you anticipate receiving notifications on vulnerabilities, reviewing and analyzing these, and addressing any issues accordingly.

Integrating all of the requirements of the IEC 81001-5-1 standard into the various standard operating procedures of your QMS is not going to happen with the snap of your fingers. The good news here is that you may not need to claim conformity with the standard as such, but instead you are free to use the standard to harden your approach to cybersecurity issues. You may end up taking a piecewise approach to the standard where you first identify it as relevant guidance, perform a gap analysis comparing your existing processes to the expectations of the standard, and only then attempt to write in the exact requirements. The requirements given by the standard do act as checklists that you can use in your software development activities.

12.6.4 ISO/IEC 27001

In the above, we mostly discussed product-related security, and did so with a definite affinity for patient safety. In this context, information security exerted most of its weight via possible harm to people, even if the immediate object in the crosshairs was information and not human beings. If we are willing to broaden our scope from the medical-device focus on patient safety and look at the management of information security on an organizational level, we may also consider the ISO/IEC 27001 standard on information security, cybersecurity, and privacy protection.

This standard does not replace our ISO 134865 standard on quality-management systems or any of the other key standards for medical devices discussed above, but it may provide elements to incorporate on the organizational level to ensure cybersecurity and information security are adequately addressed.

The point here is that a single supreme developer with an exceptional talent for producing safe, secure code will be wasted in the war on cyber threats if the whole company isn't ready to do its part. Other developers and staff acting on their own may sink the ship unless the organization itself takes a firm hold of cybersecurity and sees to the relevant resources and guidance being in place, and the staff being adequately trained for their roles in security. The call to have appropriate training programs and to demonstrate security expertise for personnel appropriate to their roles is also echoed by IEC 81001-5-1 in its Clause 4.1.4. Ideally, this training and the documentation of education, skills, and expertise should be part of your overall personnel management in your QMS (see Juuso 2022).

It is similarly important to acknowledge the role of cybersecurity and information security in the design activities for your software. It all starts with having a good security policy for the organization, having appropriate SOPs in place, and having a solid, practical organizational plan for how the various security processes are assigned at the organization. Ensuring adequate staff competence for the activities expected of them is just as important under ISO 27001 as it is under ISO 13485. Having solid, secure organizational foundations here will give you the best launchpad for creating secure software, training, or instructing your users as is appropriate, and ensuring also the long-term maintenance of your software is on a secure footing. As with the other parts of your QMS, expect cybersecurity to also be a part of your internal auditing activities going forward, even if you don't go for separate certifications according to ISO 27001.

12.7 Artificial Intelligence

Artificial intelligence and the multitude of technologies this label encompasses act as the modern-day touchstone for any

> *software-based medical device. Software that doesn't brandish some form of an AI/ML flag will seem old-fashioned. The hopes and dreams of the entire industry seem to be loaded onto this bandwagon, but it remains a subject of keen interest and much debate about what sort of a driver's license should be required of the devices at the helm.*

Great effort and considerable hype presently float around any mention of an AI-enabled medical device. Both regulators around the world and the big international standardization organizations are scratching their heads to come up with the best practices to develop, apply, maintain, and regulate AI technologies. New regulations are drafted, existing work in standardization scoured, and new thinking is published in the form of position papers, draft guidance, and white papers. To be honest, an overwhelming amount of work is presently taking place and the sheer volume of the output from this, as well as the occasionally unharmonious terminology used, is fast becoming a mountain to scale. Information overload and the problem of separating the wheat from the chaff or understanding how the different schools of thought agree or disagree, are constant issues in this fertile field.

Given the above, it is somewhat encouraging to learn that the present legislation and standards have built a solid foundation against which we can evaluate each new AI product proposed and assess its benefits and risks to patients. One of the cornerstones of this foundation is set by the Declaration of Helsinki from 1964. The declaration developed by the World Medical Association (WMA) as a statement of ethical principles for medicine involving human subjects is a brief 37-point document. The document primarily looks at medical research but ends up setting the scene for how physicians are to address the best interest of patients, especially in the context of investigating new treatments or interventions.

The layer of global legislation written on top of this declaration appears to adhere to these stipulations and objectives. The call for safe drugs has been around since at least 1906 when the Pure Food and Drug Act was passed in the US as a response to encountered fatal issues in substances available to people. In 1938 this was extended to also address medical devices, and a few decades later efficacy of the devices became a part of the picture. In Europe, the story is the same, and the advent of the new European Medical Device Regulation (EU 2017/745) was meant as a new step along this path to regulating the constantly evolving and increasingly complex devices. The great news here is that we already know what the large goals are in both safety and efficacy, and addressing the best interests of patients, even if we don't yet understand what AI technologies, best practices, and exact configuration items these correspond to. The same is true for the key medical-device standards we have today: they may not speak of artificial intelligence per se, but they show us the technology-agnostic goalposts to compare our devices and operations against.

Many organizations around the world are working on translating the general goal posts to concrete design choices and control options for artificial intelligence. Everyone and their uncles want the benefits AI can afford us, but no one wants to flirt with the risks that may also come at the same time. For this reason, it is understandable that much of the conversation in a regulatory or clinical context at the moment concentrates on the topic of transparency, i.e., understanding what it is that AI might be telling us when it is telling us something. On the highest level, this conversation can be distilled into a comparison between a black-box and a white-box approach to AI, as illustrated in Figure 12.5.

In the figure, a nihilistic example of two AI devices is given. One of the AI models is a black box, meaning it has learned some patterns by itself and we don't actually know what phenomena it bases its judgment on.

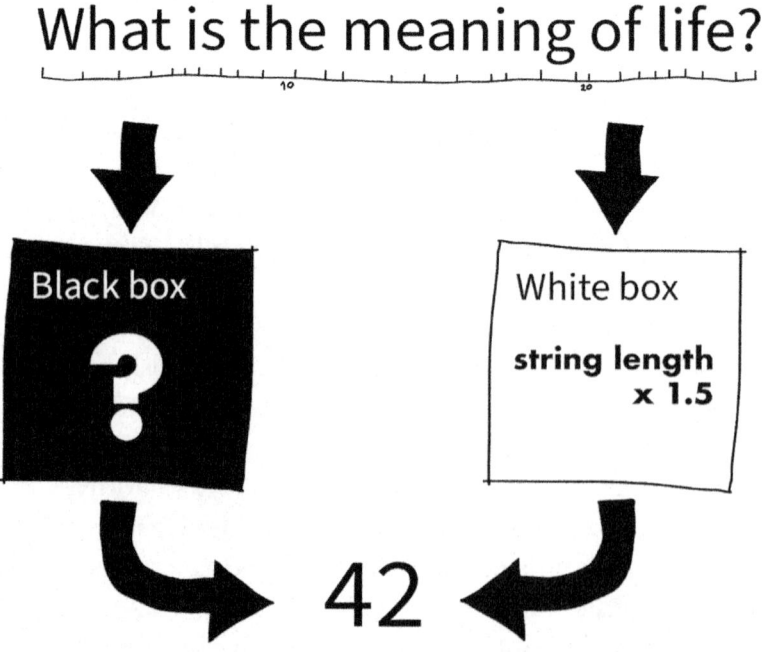

Figure 12.5 **The difference between black-box and white-box AI.**

The second AI model is a white box, indicating that we do have an understanding of how that model works. In the figure, both models give us the same answer to the meaning of life as given in the *Hitchhiker's Guide to the Galaxy*, but only with the white-box model do we understand what led to that particular answer.

The figure appears to frown upon the idea of using a black-box model, and many regulators have taken a very stern view towards such models in the past but given large amounts of high-quality data it is a feasible model for AI. Techniques also exist to lighten the black box into some shade of gray so that we could understand what is happening inside the box. At a recent FDA event, the agency gave a number of cautionary examples of using black-box models where, for example, an AI had latched onto the doctor's annotations made on radiological images to detect the phenomena already labeled by the doctor. Needless to say, the model would have been utterly useless without the doctor first doing the work. Nonetheless, AI holds great promise for the future of medical devices and the health of patients.

The work on enabling artificial intelligence applications in healthcare is rooted in the ongoing work to regulate increasingly complex software in the field of medical devices. The newly adopted European Medical Device Regulation (EU 2017/745), the EU Artificial Intelligence Act currently taking shape, and programs such as the Pre-Cert initiative run by the US Food and Drug Administration (FDA) all look at the intersection of software and healthcare with a keen eye towards the future. At the same time, cloud-based products and services are increasing their foothold in the sector, particularly in the domain of health and wellness applications where the path to user adoption is generally not hindered by considerations such as installation and maintenance inside hospital firewalls. These connected platforms will improve the availability of health data and provide a fertile field for the application of AI to tackle health issues of many sorts.

It is noteworthy that not all software is a medical device, and not all software modules in a medical device are to be considered medical in nature, either. The intended use specified for the device by the manufacturer is the key to assessing whether a piece of software, or some part of it, is a medical device or not. In Europe, the Medical Device Coordination Group (MDCG) of the European Commission has published helpful guidance on how to classify software (see MDCG 2019-11). According to the guidance, e.g., software used for the storage and transfer of medical data is not a medical device unless it alters the data somehow. Conversely, any analysis performed for a medical purpose will likely lead to designation as a medical device.

When the product or service is to be considered a medical device, a number of international standards come into play to govern the development of those devices and the organizations behind the manufacturing and distribution of the devices. Chief among the standards are the ISO 13485 standard on quality management, ISO 14971 on risk management in medical devices and the IEC 62304 standard on software life-cycle processes, all of which we have discussed in this book. These three standards form the basis of much of the work behind developing medical devices, but also other standards factor in to address clinical evaluation and usability engineering, for example. The key standards are stable at the moment, but due to the great interest in AI and other emerging technologies that are shaping the future of software in healthcare it can be expected to evolve over the next 5–10 years. If artificial intelligence plays a part in your future plans, as it is likely to do sooner or later, this is all development to watch.

12.8 Cloud Computing

The iconography for cloud services invokes vistas of perfect blue skies and bubbly cumulus clouds. An equally apt metaphor might be Spider-Man laying on an interconnected patchwork of spider webs stretched out across the sewer system of New York City, but somehow cloak-computing doesn't quite have the same ring to it. Nonetheless, clouds are here to stay every bit as much as Stan Lee's creation.

All forms of cloud computing, including the proliferation of so-called Software-as-a-Service (SaaS) tools and services, have taken over the world of computation. No one remembers the mail-order software catalogs used for software delivery in the decades past – these days a floppy disk is only really good as a coaster for the cold drink on the programmer's desk. If that. There are still critical use cases for non-cloud software, of course, but even in the world of medical computation, cloud computing has come to stay.

The cloud is used for storing data, performing calculations, and building bridges between what used to be distant silos of creating and using data across industrial domains. The cloud also scales to increasing stress from growing pools of users and the ever-new features requested in a way that puts personal workstations to shame. Adding new resources to a cloud service doesn't usually require a screwdriver and is occasionally as quick to do as the click of a mouse. It is thus not hard to see why cloud computing is an attractive proposition.

Cloud computing combined with an Internet of Things (IoT) mindset means that physical devices are mostly just sensors or drill bits pumping out data which is then streamed to the cloud for processing. The link to the cloud can reside within the sensor, or more often for now, with a mobile device, tablet, or laptop computer relaying the data from the bedside to the "bitside". All of the user's interaction with the data then takes place on the cloud.

The computational prowess available on the cloud for a fee, the plethora of available solutions for data management, and even the ready-made libraries of analysis algorithms and tools all mean the temptation to migrate to the cloud is great. There are no free lunches and even AI can be a mirage of sorts, but the promise of a plug-n-play environment for analyzing patient data and investigating the appropriate diagnoses is a tantalizing prospect. To some extent the cloud delivers here: once all of the data flows to a unified repository and queries are put at the fingertips of the users, there is little reason to doubt that new knowledge would not be distilled. The code might work the same on a personal workstation, but the data probably would not find its way there, and so the proposition of the cloud is quite different from the standalone. The potential here is limitless.

The cloud can be used as a solution to expedite software development, delivery, and adoption as long as we observe the following caveats:

- Risk management must keep up with the particulars of the new environment and the new features it enables – the old risk analysis conducted for a standalone solution does not hold water when the solution is taken to the cloud
- The physical location of the data and its transmission path matter greatly and are usually subject to several different regulations – redundancy and load distribution may have geographic dimensions that also need to be addressed
- The stacks of software components living on the cloud don't just give you neat features out of the box but also create expectations for appropriately addressing, e.g., SOUP validation and platform updates
- The user of network functionality must be addressed in the Instruction for Use (IFU) as per IEC 62304, IEC 82304-1, and the applicable regulations
- Defining what is critical performance for the device in a worst-case scenario will get to be increasingly difficult as the number of moving parts involved increases and the parts are on possibly differing update cycles – you may need to interact with the suppliers of each component

- The needs to appropriately address cybersecurity and information security are only heightened when you use the cloud to process and store your crown jewels

In addressing the above, you will find that your architectural design is a key piece of the information you'll need and also the context where you will be looking at your risk-management activities repeatedly. As you'd expect, if you are building a critical medical-device service on the cloud, even when using all the usual suspects for technology, your task will be much more complex than following the typical assembling instructions for a flatpack piece of furniture. One way of looking at the cloud is that it brings new types of interactions into the mix, and possibly opens the door to new malicious and inadvertent sources of harm. The man in the middle won't be your customer, but they might be one unwanted type of user to address in your plans.

There are two sure ways of failing in the adoption of cloud dimensions for your medical software. Firstly, you might fail by thinking that whatever worked in the standalone world would be enough. Secondly, you might be discouraged by the increased complexity and not look deep enough into what the cloud could mean for your application. Don't fall victim to either bear trap, or you'll miss out on the opportunities that cloud computing can bring.

12.9 Edge Intelligence

When Tom Cruise finds himself at the edge of tomorrow, he has a bit of a groundhog day, but ultimately cracks the code and beats the artificial intelligence at its own game. That is the vague promise we see in a technology called edge intelligence – not that it would dethrone AI, but that it might offer an interesting methodology to consider in the context of medical AI.

Edge intelligence, or Edge AI, refers to a brand of artificial intelligence where the intelligence is not located in a standalone device or in the cloud, but on the edge of the network. Many of the same concepts apply to Edge AI as to cloud computing, edge computing, and artificial intelligence. Yet, the technology has particular appeal in the context of medical devices where the dataset can be debilitatingly large, and both the data and the proprietary algorithms may need protecting – as well as the human lives connected to the algorithms.

Work here is rooted in the larger cross-disciplinary investigations into enabling AI in healthcare and is also related to the regulation of increasingly complex software in the domain of medical devices. The EU MDR, EU Artificial Intelligence Act, and programs such as FDA's Pre-Cert initiative all have a bearing here. In this context, edge intelligence could offer a novel practical solution to overcoming traditional issues faced by cloud solutions (e.g., issues of moving large amounts of sensitive data to servers outside the hospitals, or alternatively installing and maintaining cloud installations inside hospital firewalls) and stand-alone software solutions (e.g., protection of proprietary software onsite, limitations on computational power) by offering a new low-latency and high-power environment for computation with a high scalability factor. The edge-intelligence environment could offer an appealing single-use environment, the software equivalent to a cleanroom, that allows the high-volume data, the proprietary analytics algorithms, and the doctor to meet in an environment that not only minimizes data transfer requirements but also does so in an inherently cybersecure way due to its transient on-demand nature.

As a flavor of AI and an exception to the now-traditional classification of devices along an axis from standalone to cloud, Edge AI may prove to be an interesting environment to consider in designing medical software. In this book, we bring the technology up as food for thought, as an outlier perhaps challenging our notions of what it means to be online or offline.

Chapter 13

CONFORMITY ASSESSMENT

After all the hard work of developing software and deciphering the applicable requirements, the definitive test for your software may come in the form of an official IEC 62304 conformity assessment. You may think of this as the redcoats and the bluecoats assembled across a misty field at dawn, it may feel like that after all the effort expended by both sides in preparation, but the point of the assessment is to ensure your development process and its deliverables meet the expectations. Unlike at a battlefield, the expectation here is not that you will die, but the goal is that no one else will suffer either.

As remarked earlier in this book, the role of software in medical devices has grown constantly over the past few decades. Not only is it increasingly common to find some form of software in many medical devices, but also the type of functionality enabled by that software and the role of such functionality plays in forming medical diagnoses and guiding treatment is continually growing. This trend is only set to continue as increasingly sophisticated technologies, such as advanced machine learning and artificial intelligence, are brought to bear on the medical issues we face.

The result of this development is that software is increasingly the subject of scrutiny both in terms of safety and in terms of conformity to applicable requirements. Both of these are assessed against the context set by the intended use of your device, particularly if this involves the introduction of some new functionality into the inherently complex and diverse care environment. This heightened focus will also invariably lead to the addition of new software-specific requirements into both medical-device regulations and standards. Some of the requirements will speak directly to software and others will indirectly affect the development, maintenance, and use of software – or the management of the organizations involved (e.g., ISO 13485, ISO/IEC 27001).

The translation of general medical-device requirements to the specific case of software, and the interpretation of these requirements in terms relatable to software and readily understood in the context of software development, is already a challenge. The complexity of software, the distributed nature of some software, and the effects software can have on other devices over interfaces and network connections often require in-depth analysis to assess. This is particularly the case when instead of a single piece of software the whole consists of several software systems spread over multiple servers, the interaction of which may vary. This may be akin to a surgeon's scalpel not having a blade to cut with during some denial-of-service (DoS) attack, but it may also be the difference between having a dull rusty blade or a pristine sharp blade as IT enables ever newer advances to take place. With great power comes great responsibility, as one webby superhero of a comic-book movie might say. That responsibility weighs on the manufacturer, but it also weighs on the assessor.

Note that conformity with ISO 13485 and the regulatory requirements related to it are usually the topic of another conformity assessment, such as the assessment preceding CE-marking and the release of a medical device.

13.1 Requirements Placed on the Assessor

The assessment of the safety, performance, and regulatory conformity of software is often a markedly more complex task than in the case of some simple physical what-you-see-is-what-you-get medical device. In such situations, the assessor is expected to have a good command of standards and regulations but to also be in a position to interpret the requirements based on their experience with similar devices and knowledge of commonly accepted software technologies. In many cases, the assessment calls for a multidisciplinary skillset to weigh several factors at the same time and to be able to identify relevant design features for close scrutiny. Occasionally it may even be necessary to form assessment teams to tap into all of the required skillsets.

The assessment of software thus calls for the following needs to be fulfilled by the assessor or the assessment team:

- Appreciation of the regulatory requirements that are applicable to the software in question (e.g., with respect to EU MDR, harmonized standards, the concept of the State of the Art, IEC 60601-1 Clause 14, IEC 62304, and IEC 82304-1)
- Appreciation of risk-management processes and documentation which form a cornerstone of achieving the necessary level of safety and performance for a device (e.g., in respect to EU MDR, IEC 60601-1 Clause 14, and IEC 62304)
- Appreciation of software-development life cycle models (e.g., the classic waterfall model, agile models) and the deliverables produced via the individual phases of these models (e.g., requirements analysis, verifications, test cases, and review processes)
- Appreciation of software technologies including those related to programming languages, software architectures, and cybersecurity models

All of the above are naturally also a challenge for the manufacturer of the software. For each of the above aspects, the manufacturer must be able to point to an activity, a process instruction, or result document, and the acceptance criteria according to which the software may be judged to be in conformity with the relevant requirements.

The assessment of software is also impacted by the networked architecture of some of the current medical-software devices. Whether the software and its host device form a standalone device or are part of a larger network of connected devices and software each contributing to the intended use somehow are markedly different situations. Understanding all of the moving parts in the latter, and their role in the clinical environment for meeting the intended use may become quite complex. It was once enough to understand hardware and firmware, but now that is no longer necessarily the case.

In case of assessing medical-device software that is not connected to a network the following is demanded of a competent assessor:

- Ability to identify the applicable regulations and standards, and knowledge of these
- Ability to identify the appropriate assessment criteria (e.g., IEC 60601-1 Clause 14, IEC 62304), and knowledge in applying the criteria to real-world cases
- Knowledge of software-development life cycles and the deliverables involved at each stage

- Knowledge of software development and the ability to read code
- Experience in software architectures and requirements (e.g., IEC 60601-1 Clause 14.7, IEC 62304 Clause 5.2.2, and possibly even IEEE 830)
- Ability to understand documentation requirements and traceability in the context of risk control or mitigation (e.g., IEC 60601-1 Clauses 14.7, 14.10, IEC 62304 Clauses 5.2.3, 5.2.6, 7.2.2)
- Experience in working with software tools (e.g., interpreters, libraries, requirement-management software, and integrated development environments of various sorts)
- Command of essential concepts such as basic safety, essential performance, and risk control measure, and their application to software assessment
- Command of ISO 14971 risk-management principles and requirements, and practical experience in their use in assessments

If the device to be assessed is connected to a network the assessor must also meet at least the following requirements:

- Expertise in cybersecurity and information security principles, solutions, and technologies (note that several product standards already make requirements on security issues)
- Expertise in usability engineering and risk management in a complex network of human–computer interaction
- Expertise in network topology and its impact on safety and performance
- Knowledge of cloud technologies including the typical differences in their implementation (e.g., private cloud, public cloud, hybrid cloud) and the risk scenarios involved in each
- Command of a wider set of programming languages and programming tools in simultaneous use

Based on the above it is apparent that the level of complexity increases significantly as we move from the assessment of a single standalone device to a networked device consisting of diverse components each perhaps working under different parameters. As the intended behavior becomes more elaborate, also the sheer number of software components used in a device often increases. The demands placed on the assessor's command of risk management, in particular, are heightened at the same time (see IEC 60601-1 Clause 14.6 and IEC 62304 Clauses 4.2 and 7).

13.2 The Assessment

The IEC 62304 assessment may take several forms, but the objective is always the same: assess how and to what extent a manufacturer's software-development process and documentation meet the requirements of the standard. Based on our personal experience we have found two alternative models of conducting the assessment to work well in practice.

The first model is a typical audit-type assessment. Here the assessor goes through all the requirements one by one and assesses how the documentation answers to the requirements of the standard while also observing any applicable regulatory expectations. Any observations are recorded using an auditing form or the IEC 62304 test report template.

The second option is to ask the manufacturer to prefill a test report form answering the same questions and then provide this information as well as any invoked documentation to the assessor. The assessor will then go through the answers and evidence to provide the final conformity assessment. In practice, this model has often been the most cost-effective implementation for the assessment as the correct documentation is identified ahead of time by the manufacturer and a large portion, if not all, of the assessment work may be performed offsite.

The choice between the assessment model is up to the manufacturer and the assessor. Experience has shown that some form of a preliminary assessment is often a practical step to take. During this preliminary assessment the manufacturer and the assessor both get to calibrate their expectations– the manufacturer regarding what documentation is in fact needed, and the assessor regarding the manufacturer's overall setup and the status of processes and documentation. When the manufacturer's set of documentation is assessed to all be there ready and waiting, the assessment itself is begun. Occasionally the preliminary assessment may lead to the conclusion that the documentation is so woefully lacking that no assessment should in fact be performed, in which case both time and money are saved.

The tables below, Table 13.1 and Table 13.2, show excerpts from the type of test reports filled in for medical devices during IEC 60601-1 and IEC 62304 assessments, respectively.

On the test reports, the first column references the clause of the standard, the second column contains an explanation of the requirements (note that there may be multiple requirements under a single clause number), the third column describes the manufacturer's evidence for meeting the requirement, and the fourth column provides the final assessment of conformity to

Table 13.1 Excerpt From IEC 60601-1 Clause 14.6 Test Report

Clause	Requirement	Manufacturer's evidence	Pass/Fail
14.6	*Risk management process*		
14.6.1	Manufacturer has considered software- and hardware-related HAZARDS when incorporating PEMS into and IT network.	See below.	P
	The RISK MANAGEMENT FILE (RMF) includes foreseeable HAZARDS, IT network characteristics, 3rd party components, and legacy subsystems. (ISO 14971 Cl. 5.3)	The RMF references specific hazards. The required items appear in the risk analysis. Real examples are viewed from the product-specific risk analysis files.	P

Table 13.2 Excerpt From IEC 62304 Clause 5.4 Test Report

Clause	Requirement	Evidence	Pass/Fail
5.4	*Software detailed design*		
5.4.1	[B, C] The MANUFACTURER has subdivided the software until it is represented by SOFTWARE UNITS	The detailed design extends the architectural design down to the level of software units. The detailed design is given in document DOC-341144.	P
5.4.2	[C] The MANUFACTURER has documented the design with adequate detail to allow correct implementation of SOFTWARE UNITs.	The detailed design meets the expected level of detail. The detailed design is given in document DOC-341144.	P

the requirement. Note that the final assessment might also be "N/A" if the requirement is not applicable to the type of device in question.

The assessment is then done when the manufacturer has provided satisfactory responses to all requirements. If the point of the assessment is only to gauge the status of operations and identify points of improvement some of the answers may be missing or fall short of the expectations of the standard. Similarly, if the assessment is to be utilized in the notified body context to ensure appropriate operations as the basis of CE-marking a device some fails in certain requirements may be acceptable for the final device conformity assessment to still proceed. This is a topic of conversation between the manufacturer and the notified body before the assessment takes place. If, however, the objective of the assessment is to get a clean bill of health in the certification body context, and under the CB scheme (see Section 1.24), all the requirements must be passed for the assessment to go through. Even a single "fail" on a requirement will torpedo the bill of health for your software.

It is important to remember that the scope of the assessment, and the extent of required documentation, depend on the specified software safety classification (see Section 4.3). Here the ABC classification corresponds with the risk profile of the software, from low (A) to high (C). The classification is based on the risks involved, and the reasoning for a particular classification must be proved using documentation – and this should also demonstrate the incorporation of appropriate clinical expertise. Should the classification of the device change in the future, a new assessment and new evidence may be required.

After a passed assessment the assessor will provide a final test report which is then useful as a bill of health for your software and software-development processes according to the IEC 62304 standard. This report may be of great interest to your notified body for moving ahead with the conformity assessment of the device and your quality management system.

13.3 Typical Shortcomings in Assessments

Three of the most typical kinds of shortcomings encountered in IEC 62304 conformity assessments are:

- **Inadequate definition of requirements**
  (IEC 62304 Clause 5.2.2, IEC 60601-1 Clause 14.7)
- **Insufficient or too narrow risk analysis**
  (IEC 62304 Clause 7, IEC 60601-1 Clause 14.6)
- **Insufficient verification and validation activities, lack of evidence**
  (IEC 62304 Clause 5, IEC 60601-1 Clauses 14.10 and 14.11)
- **Errors in change management over time**
  (IEC 62304 Clauses 8 and 9)

Requirements elicitation is notoriously difficult and at the same time critical for the outcome of the software-development project (see Section 5.2). Taking all of the relevant inputs into consideration from the standards, regulations, and all of the involved stakeholders is no simple feat. Many software project failures can be traced back to incomplete or incorrect requirements elicitation. Thus, it is not surprising that requirements also frequently come up in IEC 62304 assessments. Misinterpretations, overly simplistic approaches, and flat-out blindness to meeting the requirements of the standard are the usual culprits here. Typical challenges here include:

- Identification of the software version each requirement applies to
- Identification of the software platform each requirement applies to
- Determination of requirements regarding essential performance (see below), usability, and risk management
- The interfaces and communication between different software systems are not covered (this creates further issues for the validity of test cases and the conformity to several IEC 62304 requirements)
- Incomplete, conflicted, or unrealistic system requirements are accepted, which causes issues for deriving software requirements
- Incomplete, conflicted, or unrealistic software requirements are accepted
- Planning documentation is insufficient and may lack a description of the software-development life cycle
- Traceability between design documentation breaks down (IEC 62304 Clauses 5.1.1c, 5.3.6, 5.4.4, 5.7.4b, 7.3.3, 8.2.4)
- The operation of the software and its operating environment has not been adequately described (this creates further issues for relying on the derived requirements and subsequent deliverables)

Risk analysis, too, is an area where manufacturers may struggle. Identifying risks may be as difficult for some manufacturers as identifying requirements, and the two are often related to one another. Both overly simplistic and overly complicated risk-classification schemes are occasionally used, and the linking of risk probabilities to real-world measurement data may be problematic. What matters here most is that the identified matrix of risks is sufficiently comprehensive, the prioritization of risks to put under the microscope is sound, and that all risk control and follow-up actions are appropriately seen to. Typical challenges here include:

- Risk management during software development is non-existent or lacking, e.g., no risk management plan is available, risk identification is done by the developer themselves in a vacuum, no risk-management expertise is present, no clinical specialist or domain expert is involved
- Information security risks are inadequately addressed, particularly in the context of network-connected or cloud-based devices

Verification and validation activities may also be a bar of soap with which to grapple for some manufacturers. Here the principal issue appears to be that not all designers understand the concepts of basic safety and essential performance, especially when it comes to determining appropriate validation criteria for software. Let's thus take a moment to discuss these concepts.

Basic safety (IEC 60601-1 Clause 3.10) refers to those structural properties of the device which are intended to prevent the device from becoming dangerous under normal conditions and so-called single-fault conditions. In the case of software, this translates to those aspects of the medical device which involve software in, for example, monitoring segregation interfaces, handling power management, and reacting to overheating issues.

Essential performance (IEC 60601-1 Clause 3.27; see Section 3.1.2) then refers to the performance of some clinical function to a level below which unacceptable risk might occur. The level may be specified as limits by the manufacturer. In practice, the clinical function is often directly linked with the intended use of the device. The collateral (IEC 60601-1-x) and particular (IEC 60601/80601-2-x) standards of the IEC 60601 family of standards place direct requirements on essential performance. These requirements may, for example, address requirements analysis, design, test-case definition, verification, validation, and risk management. For software, essential performance often translates to the ability of the employed algorithms to detect, identify, and segment some medical features from biosignals.

IEC 62304 does not reference either of the two concepts, but both are relevant, practical tools for appropriate requirements definition (IEC 62304 Clause 5.2.2) and risk analysis (IEC 62304 Clause 7).

Typical challenges here include:

- Insufficient verification plans and reports
- Insufficient validation plans and reports
- Missing validation reports (note that this is not required by IEC 62304, but is expected by regulations and standards such as IEC 60601-1, IEC 82304-1, and ISO 13485)

Change management is another typical area of *mea culpas* for software manufacturers. Once a software is finished, assessed, and taken through all the necessary regulatory hoops it may be all too easy to forget that also its maintenance is important and regulated. Every once in a while you hear developers make statements like bug fixes should have an express pathway under the radar, or small fixes don't require any impact assessment, which is why changes are occasionally inappropriately handled. The stress surrounding some urgent bug fix may also be conducive to cutting corners. Nonetheless, recording and appropriately handling proposed changes is a big part of IEC 62304 and what is, in fact, expected in medical devices. Typical challenges here include:

- Performed changes are not recorded
- Description of changes does not cover what has been changed, why it has been changed, or how conformity after the change has been assessed
- Impact of the change on documentation has not been assessed
- Changes in regulations prompting software changes have been missed

The shortcomings discussed above, and others like them, cause delays and increased expenses when discovered during conformity assessments. The worse the detected issues are the longer it will take for the manufacturer to correct them, and the longer it will be before the assessment is passed and the device can be put on the market. Issues in information security, for example, may lead to new test cases being generated, risk management being revised, accompanying documents being updated, and more likely than not also new programming work taking place. Unless a manufacturer is well-versed in IEC 62304, a pre-assessment meeting with the certification body will help prevent these issues, calibrate expectations, and set the appropriate schedule for work leading up to the conformity assessment.

13.4 What Happens Afterward?

After you have the IEC 62304 test report in your hands you can breathe a sigh of relief. The test report as well as any other testing reports and certificates you may have obtained for your product form a part of its technical documentation. This body of documentation may then be used to demonstrate the conformity of the device to applicable regulations and standards.

This documentation may be requested by the notified body and your regulatory authorities from time to time.

You will need to maintain the documentation up to date, and also observe any revision needs brought on by changes in regulations and standards, or by changes you make to the product or its intended use. It is important to remember that some of these changes may trigger a need to reassess the conformity of your software in part or in full. The regulations and any contracts you may have signed with your certification providers will specify such triggers, but when in doubt consult your notified body or a friendly regulatory consultant.

In general, undertaking IEC 62304 testing does not commit you to perform annual retesting, as is the case with maintaining ISO 13485 certificates of your QMS. However, the IEC 62304 test report will have to be redrawn if changes are made or new functionality is added beyond the previously agreed parameters. The test report is always written for a specific software version or a range of software versions allowing for certain types of controlled changes to be made (see Section 8). If you move beyond this range, a revised or completely new test report may be needed to preserve conformity. Note also that IEC 62304 Clause 6.2.5 sets requirements for making notifications of software changes (see Section 6.2). In practice, any significant change to the medical device will also interest the notified body.

Chapter 14

REGULATORY APPROVAL

You have built it and your certification body has given you their nod of approval on your hard work. There's just one more check left and that is obtaining the final regulatory approval, or clearance, for your device. This is where you tie on that bandana, strap on your test reports, and march towards that troop carrier hoping the sergeant at the door agrees to let you pass without too much groundhogery. The training montages are now over.

After jumping through all of the hoops discussed in the previous sections of this book, and now feeling confident that you meet IEC 62304 requirements, you will still probably need to obtain regulatory approvals or clearances for your device. The kids in the back seat – or perhaps the investors and your management team – might already be murmuring, "why we are not there yet"? So, why did we go through all that trouble of passing the IEC 62304 conformity assessment if it doesn't get us through the finish line?

14.1 Benefits of the IEC 62304 Test Report

At the beginning of this book, we discussed how the IEC 62304 standard does not actually give us the tools to validate your software or release the final medical device that either is the software you wrote or makes use of that piece of software. Throughout the discussion between these covers, we relied on ISO 13485, and a few other standards, to add the necessary layers on top of IEC 62304. We also discussed how for medical devices both ISO 13485 quality management (see Section 4.1) and ISO 14971 risk management (see Section 4.2) are part of the fabric, and something expected of the foundations of your software-development pipeline.

The IEC 62304 conformity assessment done by the certification body (see Section 13) and the medical-device conformity assessment done by your regulatory authorities, or the notified body, are similar and hopefully always compatible, but they do have slightly different points of view. Chief among the differences here is the assessment of the clinical performance of the device, which doesn't get much attention during the IEC 62304 assessment but is the 800-pound gorilla in the room during any medical-device assessment.

During the IEC 62304 assessment, great focus was placed on the definition of requirements, the verification activities performed, and the risk management conducted. During a regulatory assessment, all of the above still have a role to play, but now much greater emphasis is placed on, for example, the intended use, the clinical claims made, the medically sound demonstration of clinical performance, and the handling of any possible adverse effects.

The IEC 62304 assessment you have in your back pocket allows you to provide evidence of your house being in order, which is not only beneficial for signing off on the health of some required processes, but it will also allow both you and your auditor to rely on the documentation produced by those processes and your product performing as described by the documentation. Being able to quote an IEC 62304 test report will not be the same as having a Monopoly card in your pocket to thwart any software-related question that might come up during an audit, but it might just give you the sense of calm in knowing your own pockets as you head into the medical-device audit process. The folks on the other side of the table may also want to see that test report for their sense of calm in knowing your pockets.

In practice, the IEC 62304 conformity assessment is a useful and commonly expected health check of any medical device incorporating or consisting of software. In theory, the assessment can be performed by the manufacturer themselves, but demonstrating adequate competence and the required objectivity may be problematic if done this way. This is also true in the case of unaccredited or otherwise untrusted testing providers who in the eyes of the regulatory authority are unknown quantities.

If, however, the IEC 62304 test report can be trusted, the regulatory authorities can more easily accept that software development occurs in accordance with the expected standards and that the documentation produced throughout the development life cycle will meet their expectations.

Knowing this, the remaining audit can address the regulatory expectations with less ambiguity over how the requirements should be interpreted in this specific context. In other words, even if an IEC 62304 assessment is not required of you in your circumstances, you would still benefit from it by going through your activities in a more easygoing setting with genuine software experts and getting a report on it all that allows you to anticipate the sort of questions you will get regarding your software-development activities, and either feel confident in your setup or work on the detected issues before the medical-device auditors ride into town. You will be able to strike up a conversation with your IEC 62304 experts in a way that your regulatory authorities will not be allowed to interact in. The breadth of the medical-device audits will be much wider than just your software-development activities, but having an IEC 62304 assessment done first should mean that you don't need to go as deep into software during your medical-device audits as you might otherwise.

14.2 ISO 13485 Certification

As part of ensuring that the foundations for your software-development activities are solid, you will also be asked to show evidence for your overall quality management meeting the expectations set for it. Increasingly this will involve the use of the ISO 13485 standard, which was discussed briefly in Sections 2 and 4.1 of this book and covered in-depth in our previous book (Juuso 2022). This will also include ISO 14971 risk management (see Section 4.2), which is generally built into the QMS manufacturer's setup.

Whether you have already obtained certification for your QMS, will do so in concert with your first device certification, or are technically exempt from certification but still expected to adhere to the standard, it should be a part of your approach to conformity with regulatory requirements. Check your regulations on what is expected of you based on the type of devices you work with but expect ISO 13485 to come knocking sooner or later.

As discussed in Section 2, IEC 62304 mainly fits in Clause 7 of ISO 13485, but it does shade the interpretation of all requirements of the standard if you work with software products. If you have not done a separate IEC 62304 assessment you can expect the full weight of that standard to be exacted on your operations here, along the lines discussed in Sections 4 through 9 of this book. In other words, any ISO 13485 requirement will be linked to IEC 62304 requirements by the auditor and your process instructions, body of development documentation, and other records will be analyzed as evidence of meeting the expectations or falling short of them.

It is also possible to obtain a voluntary ISO 13485 certificate with a note on IEC 62304 compliance for the software-development activities. Such a certificate may be particularly practical for software subcontractors who produce software used as part of a medical device or with medical devices.

14.3 Medical-Device Conformity Assessment

After your software is ready, and it has passed the IEC 62304 conformity assessment you subjected it to (see Section 13), it is time to proceed toward that final regulatory gate with a sense of calm. In the EU, you will now be talking with your notified body. In the US, you may be looking at arranging a pre-submission meeting with the FDA or perhaps confidently putting the submission together. In other parts of the world, you will be staring at letterheads with acronyms like TGA, NMPA (formerly CFDA), Anvisa, and the like.

When inspectors from these organizations come by you can expect some poker faces and tough questions. Rightly so as the hopes and fears of every patient, doctor, and consumer out there weigh on their shoulders as much as these should be felt by your shoulders too. The one piece of advice we would part with in this context is to breathe. Breathe when they walk in, breathe when they ask you a question, and breathe while you give them the answers they ask. This way you will not only have oxygen to formulate the answers in your brain, but also the time to hear what they are actually asking for.

Assuming your auditors know of the IEC 62304 test report, and trust its source, the types of questions you can now expect are as follows.

- What is the intended use of your device?
- What is your device intended for?
- What clinical claims do you make for your device? How has its clinical performance been assessed? What clinical evidence do you have?
- How has risk management (incl. cybersecurity issues) been addressed in the development of the product? How are these issues and activities addressed over the life cycle of the product?

Your answers on all of the above will make heavy use of the technical file of your product, including the documentation created in response to IEC

62304 requirements. Many other questions regarding specific regulatory requirements and quality-management processes will also be asked, including staff competence, but in terms of software the above list represents the most frequently reoccurring topics in audits. If you have gone over the previous sections of this book methodically you should be in a great place to answer all these questions, and even occasionally politely discuss differences of opinion in the interpretation of some requirements.

14.4 Changes

We will discuss life after approvals in the next section, but here it is perhaps pertinent to point out that a regulatory approval is not a one-time thing. This has been brought up repeatedly between the covers of this book, but once you have gone through all the trouble of getting a product and a development pipeline set up according to all of the many expectations it would be silly to go about changing that machinery in any haphazard way.

The need to revise your QMS or your product may be a good sign or a bad sign. You may have identified some crack needing repairs, you may have discovered a new improved way of doing something, or it may just be that your product actually has customers and is being used so it also needs a little seeing to from time to time. Changes and popularity may, after all, go hand in hand: a product not used does not lead to feedback, but a product getting a lot of use may see a steady stream of change requests.

Regardless of why a product has been changed, the technical file of the product must reflect the changes (see Section 11). Changes in the QMS, changes in the standards and regulations, or changes in the product and its intended use all cause revisions to be considered in the documentation of the product and the processes around it. With certain types of changes, it may also become necessary to revisit the conformity assessments performed. In general, the types of changes and their impact on regulatory approval may be summarized as follows.

- **A change in the QMS**
    This should not immediately affect product conformity, but unless the change is trivial the notified body may be interested to hear about it. In response, a new assessment may be initiated, or the status assessed as part of, for example, the next scheduled surveillance audit. The change

here may also be a change in key personnel (e.g., management representative, quality manager, PRRC, risk manager).

- **A change in the key standards**

    Every once in a while, the standards do evolve, and it is your job to stay on top of upcoming changes (see Section 15). With most if not all changes you will be given a reasonable transition time to adopt the new standards or realign to meet new guidance. The adoption of a standard is generally optional, but to miss changes in the adopted standards is suspicious and potentially dangerous.

- **A change in the regulations**

    The effect of the change on the QMS, the product, and the possible other activities of the manufacturer will be assessed separately. The impact of revising any regulatory approvals will be a part of this assessment. Anyone who has been following the EU MDR transition will know that it can all be complicated, and both the timelines and effects involved will vary.

- **A change in the product**

    Your configuration management will have set up some form of version control over changes to the product (see Section 8). This scheme will have been the target of much interest from your certification providers, and you will have made promises to them on which types of changes you will ask their permission on. This scheme will now dictate how you address changes in the products, and how those changes can affect product conformity. Any time the change may touch on the clinical performance of your product you can expect some serious questions on the continued validity of your evidence.

Often the changes will have implications for reassessing product conformity. If in doubt, you can reach out to your trusted consultants or even your regulatory authorities despite popular wisdom suggesting that it may be easier to ask for forgiveness rather than permission. After all, you will be more easily forgiven for asking an unnecessary question repeatedly than given permission to do as you will. In practice, your QMS processes will instruct what changes are to be communicated and how, and these will have been audited by your overseers.

Chapter 15

BUSINESS AS USUAL

You built it and they came? Your product is now flying off the shelf and you are busy coming up with the next cool version that will again totally revolutionize your target market. But what should you be keeping your eye on, and what should you be doing now that it is all built for the time being?

When you first have a big new idea for a medical-software product, you probably have a gauntlet to run to get it funded, built, approved, and into the hands of paying customers. The long-term support and maintenance of that software is probably a distant blip on your radar, something you will "of course" see to once you get there and have made it big. The medical-device standards and regulations, however, want to see you well-prepared for that long game, and for you to have built that bridge before you get to it. The long game, for them, doesn't end until your software is safely decommissioned, and even then, you will be required to keep the lights on for several more years to come in case issues pop up.

The standards and regulations speak of production information, maintenance, and – most of all – post-market surveillance. The bottom line in all this work is not your financial bottom line, but customer satisfaction, patient safety, and whatever else your intended use and the regulations may prescribe. All of this is a far cry from the fail-fast and mutate-to-win mindset that may get you far in software development in other domains, such as the development of mobile games. In medical-device software, the concept of a customer relationship involves a much bigger commitment than just picking out car keys at a swingers' party – or doing your initial elevator pitch at some free-for-all investor event. In medical devices, you may not have

DOI: 10.4324/9781003454830-15

to commit to putting the offspring through college, but you may need to at least know about the future educational update needs of your software and its users. The long play is the end game.

15.1 The Everyday Processes to Now Run

Colorful language aside, the long-term existence of your software should be on your mind when you design, develop, and release medical software. You will want to, at least, address the following before placing software on the market and develop the implicated processes, plans, and arrangements as appropriate. The considerations for "business as usual" after the release of medical software thus include, for example, the following:

- Production monitoring
- Maintenance
- Post-market surveillance
- Regulatory reporting
- Regulatory monitoring (incl. the use of international standards)

Production monitoring is primarily a topic for churning out copies of a physical product over a long period of time. Production line equipment may experience wear and tear, the quality and other characteristics of raw materials may fluctuate, and you may introduce other changes to the printing press which may affect the conformity of the products you make. For software, such fluctuation may be nonexistent, but nonetheless, components such as SOUP modules and changes to the software platform you deploy your product on will be on your mind as time goes by.

Maintenance is going to be a big factor for software-based products. As mentioned in the preface to this book, software without updates will get to be an oddity over any extended period of time. It will depend on the issues you encounter, the features you want to revise or add, the nature of your product, and how soon you will be putting out maintenance updates. You have already made instructions and plans for maintenance activities before releasing your product, so the first thing to see to here is that you live by those promises.

Post-market surveillance is an increasingly important part of life after product release. You must keep an eye on how your product – and any similar devices released by you or your competitors – is fairing out there in the real world, how the science and technology around your device is evolving, and how the regulatory and standardization landscape is changing. It all starts with sales and use figures, but your field of vision should be wider here than just the mail inbox you might receive part of the feedback through. Also, remember that ISO 13485 requires you to look at servicing information as a source of customer feedback and complaints even when your customer is only indirectly involved.

Regulatory reporting may also be an activity you need to have running constantly or jumping into action at specific times or after specific events. You may need to report on your post-market surveillance activities periodically. Similarly, some of the issues you uncover during your post-market activities, or otherwise, may warrant reporting to your notified body and regulatory authorities. The timeline here will depend on the class of medical device you have, the type of event in question, and the regulatory requirements applicable to you in the target jurisdictions.

Finally, regulatory monitoring and the use of international standards will be a domain to keep a watchful eye over in the coming years. It may be that not much will change in the first few months after your software is released, but over time significant changes will occur and it is important to stay on top of such change. You will see official guidance on legislation and standards being published all the time, new regulations and standards will appear gradually, and the state-of-the-art – which in medical devices doesn't even refer to the cutting edge, only to the generally accepted good practice – will be constantly evolving. The only constant in life is change, as countless wise people have no doubt said over the last millennia.

It is, of course, impossible to keep track of everything at the same time. Our recommendation is to get involved in industry forums, follow relevant industry and scientific publishing to whatever extent you can, and perhaps utilize a good regulatory intelligence service. A good consultant will also be able to help you here.

IEC 62304 discusses many activities relevant during the post-market phase of your product (e.g., change of management and problem resolution), but many of the other ongoing activities you will need to run around your software are instructed by ISO 13485. You can find an in-depth discussion of these in our first book *Developing an ISO 13485-Certified Quality Management System* (Juuso 2022).

15.2 The Future of IEC 62304

The IEC 62304 standard is for now a way of life for medical software. The standard has been declared stable for the next few years, and it comes with such a following that it would not just be foolish, but dangerous, to feed it to a shredder without first mapping out the way forward. The needs addressed in the standard through its 9 clauses of requirements and administered through its risk-based software safety classification are very real and are not rendered obsolete by the newer challenges of agile development, artificial intelligence, and whatever else soup du jour the standard is dipped in. The same requirements will still need answers in the future.

The reality is that challenges have been identified relating to the use of the standard in the environment of today. Even with a standard this old, smart guidance tying it to modern needs and clarifying the requirements in a way that doesn't just rely on the assessor's subjective assessment across the board is clearly needed. Many of these challenges have led to the introduction of guidance or standards such as those discussed in Section 12. These documents can be used together with the aging IEC 62304 standard to give it a new lease on life. In this book, we have sketched out a joint existence for IEC 62304 and ISO 13485, both of which set up the most significant blocks of the Stonehenge that is medical-software development. In doing so we have shown that the two really are not at odds with each other, can coexist, and can be utilized to create a better environment for software development and for the long-term supply of medical-device software.

In the coming years, the form the IEC 62304 standard takes may even see dramatic turns, but whatever the metamorphosis, the DNA of the standard will still be recognizably IEC 62304 – or the standardization organizations may have missed their goal of fostering systematic, consistent, and predictable processes. Looking to the darkness of the clouds ahead, Sarah Connor scratched out the words "No fate" on that wooden tabletop in *Terminator 2*. In terms of IEC 62304 that may also be true, but the standard is still state-of-the-art, and as such, is as necessary as the aging T-800 in that film. To shred the IEC 62304 framework and leave the present adopters behind is not possible, smart, or safe.

Should the end of the world come, though, you should be well-positioned as this book has guided you toward meeting first the requirements of ISO 13485 quality management, and from this foundation, meeting the present-day requirements for software development. In the future, you can look for guidance and early warning signs of how the requirements around both of these evolve, but at present no change is imminent, and no change will happen overnight unless the values of standardization itself are put to the shredder.

15.3 The Future of ISO 13485

The ISO 13485 standard is for all intents and purposes a bedrock standard. One should never say never, but in the present stage of its use and budding adoption across the globe, it should be more likely that dinosaurs return to the Earth than the standard is reworked in any upsetting way.

We take an active role in following and taking part in the development of the standard and have done so for a number of years. As committee members on the ISO technical committee and its national mirror groups, we have watched with delight how this standard is living up to its moniker as an international standard. The challenges put to the standard are largely the same as those put to IEC 62304, but no one in their right mind wants to rock the stability and trans-continental compatibility found in the standard at the moment. A wealth of related guidance is seen as beneficial, and this will come out in due course in one form or another, but no reshuffling or seismic changes to the standard itself are to be expected. In this book we chose ISO 13485 as the backdrop to everything, and with good reason. The standard is the foundation of medical device development and realization.

15.4 The Joy of Compliance

The standards are written by experts wanting to point out the important facets of doing something, and all the various information sources and factors you will want to consider while climbing toward that elusive peak of perfection. In this book, we have been careful not to make this sound like you must know everything to make a start with anything. We have, on many occasions, pointed out things, both great and small, that you will want to at least consider on your climb. At the same time, we have wanted to distill each of the requirements to their core and show you the big motivations and driving considerations behind them. We hope we have succeeded in this and shown you the axes around which many of the moving parts revolve, and around which you can expect all future additions to also revolve.

As changes emerge, you will face transition activities of different magnitudes. Many changes in the regulations and standards may feel like they only call for minor revisions to be made in your documentation. This may even be the case for some changes. In the case of other changes, you will need to rethink some subprocess, or add some activity, and revise your instructions and documents in turn. In a more dire situation, you might find that the clinical evidence for your legacy product is called into question or deemed insufficient due to changing regulations, in which case conducting new testing, gearing up risk-management activities, and obtaining more expansive data on your clinical performance will perhaps get to be an expensive proposition. The better you understand why you have been asked to shape a process a certain way, perform some testing or risk-management activity, or provide some particular piece of documentation the better you will be prepared for reacting to – or perhaps even anticipating – some change in the external requirements applicable to your products and organization.

Shaping your operations according to commonly accepted international standards, such as those discussed in this book, will not only give you a solid, defensible way of developing software but also the most straightforward platform for addressing any future challenges. In the end, regulatory compliance will be key but maintaining compliance with the right standards will be your best bet to also remaining in compliance with the applicable regulations both domestically and internationally too.

Chapter 16

CONCLUSIONS

At the climax of the fourth Indiana Jones film, it all ends in a vortex of all the knowledge mankind has ever accumulated – and this is then comically flushed down into the depths of the jungle. We hope this has not been, and now is not, your experience of trying to fuse together the expectations of medical-software development. If it is, try to count the number of sprint cycles the fulcrum goes through in that scene, and after recovering from that distraction, reconsider.

Reading the history of medical devices, you will come across inspirational stories from the recent past that identified needs in the treatment of patients, coupled these with the necessary engineering skills, and made the world an undeniably better place. In these stories, the spark for a new medical device may have come from a fortuitous meeting of minds, like today, but the path to testing the prototypes out with real patients was occasionally significantly shorter than that expected today – perhaps only a matter of days from the initial concept. In this regard, no example is perhaps as inspiring as the origin story of the Minnesota-based medical-device mammoth Medtronic. It was in 1957 when a University of Minnesota heart surgeon, Dr C. Walton Lillehei, and an enterprising electrical engineer, Earl Bakken, dreamt up a wearable transistorized pacemaker and only days after attached a prototype to the first young patient. A year later the device was on the market, and decades later an implantable version of this same pacemaker is now commonplace.

Today this timeline may sound fantastical, and certainly, the risk-management activities involved and the investigational procedures called for are now more evolved, but the drive to meet new unmet needs and to also better meet those previously addressed is forever present. Those needs are increasingly being met by software – software embedded in medical devices and software acting as medical devices. Just as the invention of the transistor, and more specifically the use of those transistors in a metronome circuit, allowed the creation of the first heart pacemaker, the ongoing surge of software holds the promise of ushering in the next generations of medical devices. Not all software has to be dubbed "artificial intelligence" for it to change the world in some discernible increments, either.

It is, therefore, of great importance that the development and maintenance of that software can take place with as little friction and overhead as possible to support innovation and patient health. There are vastly more software developers out there who could contribute to medical software than those software developers who are willing to learn and live by medical-device standards and regulations. That is not to say that we should do away with standards and regulations, most certainly we should not. It does, however, suggest that we should make it as simple as possible to know what is good practice and what is required in the field so that the platform for developing – and maintaining – medical software makes as much sense as possible. This also has implications in academia where a great number of research projects are set up, but the fruits of those endeavors may rot away after the project ends because it would be too much work to redo everything according to what a medical project launch would require. Knowing enough about the science or the clinical setting may not be a problem in such a context, but recording that information and considering enough of the potential big showstoppers from standards and regulations may drop a wrench-shaped anchor on the proceedings.

It was this drive to figure out the requirements and the best practices and to engineer them into some cohesive, practical framework that drove our efforts to write this book. In the discussion within these covers, we have attempted to align the two major standards, ISO 13485 and IEC 62304, in a practical and purpose-driven way. That purpose was to meet all of the universal expectations placed on medical software in a way that also makes sense for the developers. We firmly believe that understanding the reasoning behind such expectations and requirements does not only mean that the demands are met, but also met better, safer, and with less friction than what could be expected if they were viewed just as something required, for whatever reason.

Writing this book was almost a two-year project that involved diving deep into the two principal standards, and a good number of their brethren, as well as the applicable regulatory requirements that these occasionally reference. We spent countless hours in discussion on the various aspects involved and reached out to other expert voices on many of the topics. In the discussion here we hope we have given a good overview of the combined requirements, identified the various linkages, and pointed out all of the key design considerations and questions you may want to address in developing and streamlining your software activities.

The discussion here in this book is the most developed and most comprehensively considered approach to a consolidated, unified model of medical software development we are aware of. Still, the solutions offered here are based on our own experience and insight into the development and use of international standards in medical devices. Your own solutions may differ in places and still make as much sense as our suggestions here. Considering your position on our suggestion will, however, give you both new ideas and the support you want in your choices. As we pointed out at the beginning of this book, there are more needles than just one in that haystack, but we hope that by reading the discussion here you will have gained confidence in knowing that haystack.

The last thought we want to leave you with here is that, if you are already accustomed to developing software on some other domain, you shouldn't feel like you have to reinvent everything for the medical-device world. If you have a solid development process that has worked for you well in the past, the chances are that it – or many elements from it at least – will also work for you in medical devices. ISO 13485 leaves great room for you to develop software between its clauses on D&D inputs and outputs. IEC 62304 adds many more markers to hit in doing so, but even it doesn't micromanage your day-to-day. In practice, just about all of the best practices and software development models you may be accustomed to from other domains and disciplines will also be translatable to medical-software development – but translation is almost certainly required.

The fundamental requirements for the medical-software development process are that all work occurs in a planned-out fashion, outcomes are approved by appropriate functions, the whole is somehow approved as well as the incremental parts, long-term maintenance is part of the big picture, effects are analyzed before executing changes, and throughout all records are maintained as appropriate. Any development model that meets these fundamentals can be mapped to the requirements of the standards and regulations. In practice, the reality of designing an appropriate development process is, of course, a little bit more complex than just fulfilling this handful of requirements, as this book has perhaps shown, but the grandest of strokes are accurately set by them. Use this realization to look at where you are today, and then think carefully about where you want to be and you should get there. This is what Deming would also advise you to do.

Similarly, the world never stops. There will be new standards and regulations to comply with, some perhaps prompted by new advanced technologies and others by increased calls to interact with your users over the life cycle of your software. Although the requirements will evolve, the fundamental needs behind each requirement will remain much more stable. We will always be asked for better, cheaper, and more powerful medical devices providing the greatest of benefits without any risks to anyone. There is no such thing as future-proof, of course, but knowing the fundamental goals behind each essential requirement, and ensuring you have lean processes in place for meeting these goals, will give you the best platform for any future developments too.

That's it for now. We searched high and low for the different kinds of needles in the haystack, and hopefully gave you new ideas and perspectives on developing medical software in the context of ISO 13485 quality management. If you agree or disagree with any of the discussion in this book, or have further comments, questions, anecdotes, or other needles from the haystacks to share please contact us. You can reach us both via LinkedIn and also by writing to sw@theqmsbook.com.

Thank you for reading our book! Like James Bond, we may return.

References

Juuso, I. (2022). *Developing an ISO 13485-Certified Quality Management System – An Implementation Guide for the Medical-Device Industry.* Routledge, New York.

MDCG. (2019-11). Guidance on Qualification and Classification of Software in Regulation (EU) 2017/745 – MDR and Regulation (EU) 2017/746 – IVDR (2019). Medical Device Coordination Group.

Pure Food and Drug Act. (1906). United States Statutes at Large (59th Cong., Sess. I, Chp. 3915, p. 768-772; cited as 34 Stat. 768).

Index

A

ABC classification, 11, 12, 92, 99, 310
Agile software development
    advantage, 264
    ChatGPT, 262
Analyze data, trends
    IEC 62304 standard, 236
    ISO 13485 standard, 235–236
    synthesis, 236–237
Applicable regulations, 20–22
Architectural design, 132–133
    IEC 62304, 134–136
    ISO 13485, 133–134
    synthesis, 137–138
Artificial intelligence, 16, 25, 213, 286, 290, 296–300, 328
Audit package, 259

B

B2B customers, 27, 32
Business as usual
    compliance, 325–326
    everyday processes, 322–323
    IEC 62304 standard, 324–325
    ISO 13485 standard, 325

C

CAPA, see Corrective and preventive actions
CB, see Certification body
Center for Devices and Radiological Health (CDRH), 56
CER, see Clinical evaluation report
Certification body (CB), 5, 52, 61, 75, 310, 313, 316
Change control
    IEC 62304 standard, 216–217
    ISO 13485 standard, 215–216
    synthesis, 217–218
    use, 231–233
Clinical evaluation report (CER), 167, 265
Cloudbased services, 260
Cloud computing, 300–302
Code *vs.* documentation, 26–27
Configuration management
    aim, 206
    A-level change, 210
    B-level change, 210
    C-level change, 210
    databases, 213–214
    D-level change, 210
    identification, 207–208
    IEC 62304 standard, 208–209
    ISO 13485 standard, 208
    synthesis, 209
    tracking, 211–213
    version numbering, 210–211
Configuration status accounting, 218, 219
Conformity assessment
    assessor/assessment team, 305–307
    documentation, 313–314
    IEC 60601-1, 309

IEC 62304 assessment, 308–310
regulatory approval, 318–319
shortcomings, 310–313
Controlled items, history
IEC 62304 standard, 219
ISO 13485 standard, 219
synthesis, 219
Corrective and preventive actions (CAPA), 56, 223–228, 231, 232
Cybersecurity, 290–291
IEC 81001-5-1 standard, 294–295
ISO/IEC 27001 standard, 295–296
jurisdictions, 293–294
product life cycle, 291–293

D

D&D, *see* Design and development
D&D deliverables, 47
D&D input stage, 158, 165
D&D output stage, 158, 162, 165
Denial-of-service (DoS), 305
Design and development (D&D), 13, 33, 57, 106, 126, 207, 224, 254
Design input review (DIR), 29, 31–33, 131, 257
Design output review (DOR), 33, 159, 166–167, 249, 257
Design readiness review (DRR), 33
Design transfer review (DTR), 33–34, 161, 167–170, 258
Detailed design
architectural map, 138
class B software, 139
IEC 62304 standard, 140
ISO 13485 standard, 139
synthesis, 141
Development life cycle, 15, 60, 105, 113, 120, 248, 265–267
Development planning
aspects, 112–115
configuration management, 115–116
IEC 62304
planning documentation, 110–111
process documentation, 111–112
ISO 13485
planning documentation, 110–111
process documentation, 109–110

synthesis
plans, 122–123
SDP, 119–121
SOP, 117–118
system development plan, 118–119
use of systems, 123–124
Development records, 256–258
Device History Record (DHR), 254
Device Master Record (DMR), 254
DHR, *see* Device History Record
DIR, *see* Design input review
DMR, *see* Device Master Record
Documentation Level rating, 12
DOR, *see* Design output review
DoS, *see* Denial-of-service
DRR, *see* Design Readiness Review
DTR, *see* Design Transfer Review

E

Edge AI, *see* Edge intelligence
Edge intelligence, 261, 302–303
Electronic health-record systems, 9
External risk controls, 96, 193–194, 277

F

Failure modes and effects analysis (FMEA), 98, 203
hazards, 281
ISO 14971 standard, 283
risk score, 282
RPN risk scores, 283
types of risks, 280–281
Fault tree analysis (FTA), 284
FMEA, *see* Failure modes and effects analysis
Foundational requirements, 89, 207
FTA, *see* Fault tree analysis

G

Gap analysis
closure activities, 103–104
rationale, 104
Global medical-device ecosystem, 56
Green-coding practices, 94

H

Hazardous situation, 94, 194–197
   IEC 62304 standard, 197–198
   ISO 13485 standard, 197
   synthesis, 199

I

IEC, *see* International Electrotechnical Committee
IEC 60601-1, 9, 37, 62–66, 86, 128, 277, 306
IEC 62304 standard, 6–7, 10, 14, 17
   activity, 71
   alignment, 107
   analyze data, trends, 236
   architectural design, 134–136
   business as usual, 324–325
   change control, 216–217
   conducting reviews in D&D stages, 170
   configuration management, 208–209
   controlled items, history, 219
   D&D validation, 172–174
   D&D verification, 170–172
   detailed design, 140
   developing, 22–23
   development planning
      planning documentation, 110–111
      process documentation, 111–112
   documentation set, 19–20
   external risk controls, 96–97
   final remarks, 99–100
   foreword, 69–70
   future, 67–68
   general requirements, 79–80
   guidance on the provisions, 85–88
   harm in relation to people, 193
   hazardous situation, 94, 197–198
   implementation, 88
   integration testing, 148–151
   introduction, 70–71
   legacy software, 48–49
   maintain records, 234
   mode of execution, 74
   modern times, 65–67
   modification implementation, 186–187
   nature of the software, 74
   non-binding notes, 75
   normative references, 76–77
   primordial ooze, 62–65
   probability, 97–98
   problem and modification analysis, 182–184
   problem investigation, 227–228
   problem reports, 225
   process, 71
   product maintenance, 188
   QMS, 90
   release
      D&D Output Review, 162
      production and service provision, 164–165
      validation, 163–164
      verification, 163
   relevant parties, advise, 230
   requirements, 84–85
   requirements analysis
      finalizing requirements, 129–130
      inputs for requirements, 126–129
   risk-control measures
      software changes, 202
      verification, 200–201
   risk management, 91–92
   safety classification, 92–94
   scope, 73–76
   software configuration management process, 82–83
   software development process, 80
   software maintenance plan, 179
   software maintenance process, 80–82
   software problem resolution, 239–240
   software problem resolution process, 83–84
   software risk-management process, 82
   software-system testing, 154–156
   software will fail, 95–96
   specific process structure, 18
   task, 71–73
   terms and definitions
      software item, 78
      software system, 77–78

software unit, 78
system, 77
test documentation contents, 242
testing, 36
type of delivery, 74
type of storage, 74
unit implementation and verification, 143–145
verification and testing activities, 107
word cloud, 74, 76, 80, 83
IEC 81001-5-1 standard, 294–295
IEC 82304-1, 267–269
post-market activities, 275–276
product documentation, 272–275
product requirements, 269–270
software development process, 271
validation, 271–272
IEEE 610.12:1990 standard, 14
Integration testing
IEC 62304 standard, 148–151
ISO 13485 standard, 147–148
synthesis, 151–152
International Electrotechnical Committee (IEC), 63, 69
ISO/IEC 27001 standard, 295–296
ISO 13485 standard, 7, 10, 16, 17
alignment, 107
analyze data, trends, 235–236
architectural design, 133–134
business as usual, 325
change control, 215–216
configuration management, 208
controlled items, history, 219
detailed design, 139
development planning
planning documentation, 110–111
process documentation, 109–110
hazardous situation, 197
integration testing, 147–148
maintain records, 234
modification implementation, 186
problem and modification analysis, 182
problem investigation, 227
problem reports, 224–225
product realization, 57–60
quality management system, 56–57

release
D&D Output Review, 158–159
production and service provision, 161–162
validation, 161
verification, 160–161
relevant parties, advise, 330
requirements analysis
finalizing requirements, 126–127
inputs for requirements, 126
risk-control measures
software changes, 201–202
verification, 200
software maintenance plan, 177–179
software problem resolution, 238
software-system testing, 153–154
test documentation contents, 241–242
testing, 36
unit implementation and verification, 142–143
vs. IEC 62304, 59–60
ISO 14971 standard, 7, 50, 78, 283

L

Legacy concept, 65
Legacy software, 48–49
gap analysis, 101–104
risk-management activities, 100–101
Level of concern (LoC), 12, 76, 99

M

Maintain records, 126, 223, 233–235
MDCG, *see* Medical Device Coordination Group
MDR, *see* Medical Device Regulation
MDSAP, *see* Medical Device Single Audit Program
Medical Device Coordination Group (MDCG), 9, 293, 299
Medical Device Regulation (MDR), 9, 66–68, 254, 265, 297, 304
Medical Device Single Audit Program (MDSAP), 56
Microchips, 62

Minnesota-based medical-device, 327
Modification implementation
    IEC 62304 standard, 186–187
    ISO 13485 standard, 186
    problem-resolution process, 185
    synthesis, 187–188

N

National certification body (NCB), 52
Non-cloud software, 300
Not applicable (N/A), 39

O

Object-oriented programming language, 13

P

PEMS, *see* Programmable medical electrical systems
PESS, *see* Programmable electronic subsystems
Plan-Do-Check-Act model, 31
Post-production activity, 191
Pre-Cert program set, 16
Prerequisites, 89
Problem and modification analysis, 180–182
    IEC 62304 standard, 182–184
    ISO 13485 standard, 182
    synthesis, 184–185
Problem investigation
    IEC 62304 standard, 227–228
    ISO 13485 standard, 227
    synthesis, 228–229
Problem reports, 221–224
    IEC 62304 standard, 225
    ISO 13485 standard, 224–225
    synthesis, 225–226
Process documents, 256
Production activity, 191
Programmable electronic subsystems (PESS), 64
Programmable medical electrical systems (PEMS), 64

Q

QMS, *see* Quality management system; Quality management systems
Quality management system (QMS), 17
    best ideas, 251
    execute plans, 250
    planning documentation, 250
    realization processes, 247–249
    release activities, 251
Quality System Record, 254

R

Regulation
    applicable, 20–22
    maintaining trust, 24–25
Regulatory approval
    changes, 319–320
    IEC 62304 test report, 315–317
    ISO 13485 certification, 317–318
    medical-device conformity assessment, 318–319
Release
    IEC 62304 standard
        D&D output review, 162
        production and service provision, 164–165
        validation, 163–164
        verification, 163
    ISO 13485 standard
        D&D output review, 158–159
        production and service provision, 161–162
        validation, 161
        verification, 160–161
    synthesis, 165–166
        DOR, 166–167
        DTR, 167–170
        validation report, 167
Relevant parties, advise
    IEC 62304 standard, 230
    ISO 13485 standard, 330
    synthesis, 230–231
Requirements analysis, 124–125
    IEC 62304
        finalizing requirements, 129–130

inputs for requirements, 126–129
ISO 13485
   finalizing requirements, 126–127
   inputs for requirements, 126
synthesis
   requirements elicitation, 130–131
   requirements review, 131–132
Residual risk, 190–191, 272, 283
Risk assessment, 130, 132, 168, 190, 194, 213, 229, 269, 289
Risk control, 190
Risk-control measures
   IEC 62304 standard, 199–200
   ISO 13485 standard, 199
   software changes
      IEC 62304 standard, 202
      ISO 13485 standard, 201–202
      synthesis, 202
   synthesis, 200
   verification
      IEC 62304 standard, 200–201
      ISO 13485 standard, 200
      synthesis, 201

S

SaaS, *see* Software-as-a-Service
SaMD, *see* Software as a Medical Device
SDP, *see* Software development plan
Segregation, 276–280
SiMD, *see* Software in a Medical Device
Single fault conditions, 65, 312
Software architecture, 134, 276–280
Software as a Medical Device (SaMD), 9, 12, 13
Software-as-a-Service (SaaS), 300
Software-development
   CB, 51–53
   configuration items, 46–48
   continuous improvement, 43–45
   controlling changes, 46–48
   evaluation, 29
   fail, 40
   forward momentum, 41–43
   life cycle, 14–16
   not applicable (N/A), 39
   pass, 40
   review, 29, 31–33

risk management, 49–51
safety, 53–54
shortcomings, 41
testing, 34–38
validation, 30–31
verification, 29
Software development plan (SDP), 112, 119–121, 246
Software in a Medical Device (SiMD), 9, 13
Software items, 13, 14, 78
Software maintenance plan
   IEC 62304 standard, 179
   ISO 13485 standard, 177–179
   synthesis, 179–180
Software of Unknown Provenance (SOUP), 45–46
Software problem resolution
   IEC 62304 standard, 239–240
   ISO 13485 standard, 238
   synthesis, 240–241
Software safety classification, 92, 192–193
Software-system testing, 152–153
   IEC 62304 standard, 154–156
   ISO 13485 standard, 153–154
   synthesis, 156
Software unit, 13
Standard operating procedure (SOP), 117–118, 246
State of the art, 7
Stockholm syndrome, 55
System begets, 12

T

Technical documentation, 187, 252–255, 289, 313
   design and development files, 254
   medical-device family, 253
Terminological reductionism, 17
Test documentation contents
   IEC 62304 standard, 242
   ISO 13485 standard, 241–242
   synthesis, 242–243
Three-level safety classification, 11
Top events, 284
Transfer Readiness Review (TRR), 33
TRR, *see* Transfer Readiness Review

U

Unit implementation and verification
  designing, 142
  IEC 62304 standard, 143–145
  ISO 13485 standard, 142–143
  synthesis, 145–146
Usability engineering, 255, 261, 273, 284–290, 292, 293, 300, 307

W

Waterfall model, 25, 107, 167, 248, 263, 264, 306

Z

Z-annexes, 7
Zero emissions goals, 94

Printed in Great Britain
by Amazon